THE WINDMILL
yesterday and today

Mr de Little, who lives in Sussex, is one of
the growing band of enthusiasts who is
interested both in the history of windmills
(of which about ten thousand stood in our
land little more than a century ago) and
in their working restoration. Now a
professional man, he at one time worked
for a firm of millwrights concerned with
the Shipley, Sussex, and Meopham, Kent,
windmills. He has been directly interested
in windmills for fifteen years, has visited a
a great many mills, assisted in their
reconstruction, and has also contructed
third-size mill models.

WINGHAM, KENT Now a capless tower, the sails have been erected on Sandwich smock mill which has been restored.

WILLINGHAM, CAMBRIDGESHIRE The last smock mill to work in the county. Ceased to work in the early 1960s.

THE WINDMILL
yesterday and today

R. J. DE LITTLE

Illustrated by 101 monochrome plates

4, 5 & 6 SOHO SQUARE LONDON W1

© 1972 R. J. DE LITTLE
Published in 1972 by
JOHN BAKER (PUBLISHERS) LTD
4, 5 & 6 Soho Square,
London WIV 6AD
ISBN 0 212 98397 0

Printed in Great Britain by
HOLLEN STREET PRESS LTD AT SLOUGH, BUCKS

CONTENTS

ILLUSTRATIONS

Italic numbers are page references.

1 · The origin of the English windmill

There is no known inventor of the windmill as we know it. The first definite records in this country date from around the year 1190, and the illustrations of this period show a very small and simple version of the post windmill, which was the basic type for many hundreds of years. Several different views are held as to how windmills first came about in this country: that they were first seen in the Middle East by the Crusaders; that they were discovered in China; that they were invented in Europe or grew up independently in many parts of the world.

The European idea seems the most likely, and certainly windmills are very widespread throughout this area. From 1190 onwards windmills spread throughout England and even, in very limited numbers, into Scotland and Wales. Their numbers increased until around 1840 or so; there were probably about 10,000; William Coles-Finch in his book *Watermills and Windmills* (published 1933) traces over four hundred in Kent alone.

DISTRIBUTION

Water power is more stable and cheaper to harness than the wind. For this reason, where there is adequate water for the operation of mills we do not find windmills. The dry Eastern counties are the great windmilling areas; Sussex, Surrey, Kent, Essex, Suffolk, Norfolk, Lincolnshire and the East Midlands, with another area around Anglesey, Cheshire and Lancashire. Areas such as Cornwall, where there are many streams, had very few windmills indeed.

DEVELOPMENT AND DECLINE

While most European countries developed their windmills to a high state of efficiency, the absolute pinnacle of design was probably reached in England. This peak probably coincided with the introduction of steam mills at the various ports. These mills ground foreign grain in roller mills to a standard of whiteness unobtainable in a windmill using stones. The public has for centuries yearned for whiter and whiter bread, culminating in the light rubber-like substance found today. Thus public taste pushed out the windmill around the big towns.

Windmills continued to produce flour in the country districts, however, until the introduction of motor transport, which enabled roller-ground flour to be transported quickly and cheaply to all areas. Many windmills continued by producing cattle-feed, but even this trade was gradually taken by the big mills in the towns.

From 1895 windmills went out of use and by 1912, when motor transport was really coming into its own, they were being destroyed at an incredible rate. World War I caused many more to close down, through shortage of timber and economic reasons. By 1930 people were realising what had happened, and some preservation was undertaken. That excellent body The Society for the Protection of Ancient Buildings enabled several mills to remain at work, and even had the mill at Stanton in Suffolk (Fig. 49) restored to work again. Alas, World War II intervened, the millwright went away, and the mill never started.

The shortage of timber again caused many mills to stop. The beginning of that fascinating period the 1950s saw about a score of mills still working, and while this number steadily diminished, a number of public bodies and county councils were concerning themselves with the external preservation of local windmills. At the beginning of the 1960s about eight mills were at work, and there was a great increase in the desire to preserve those windmills which remained in a fairly complete state. These were mostly 'shell' restorations, and much fine machinery was regrettably destroyed. The period was marred by the sad neglect of several fine mills which collapsed or were demolished. The smock mill at Outwood, Surrey (Fig. 69), was one of these and, although rather recent, was quite exceptional.

RESTORATION

The years 1965 to 1970 have been outstanding, for several complete rebuilds have taken place, and at last fidelity to original design is becoming of interest to the restorers. Groups of enthusiasts are working together to undertake restoration on mills which were considered beyond repair a few years ago. The increase in enthusiasm may be gauged by the fact that at the time of writing only about four mills are still grinding in connection with their owners' business, whereas at least five are worked quite regularly for demonstration to the public, or for the enjoyment of enthusiastic owners.

1. (*Opposite*) CHILLENDEN MILL, KENT An open-trestle post mill which was turned to face the wind by means of a tailpole. This mill is not as old as it seems and was in fact built in 1868. The mill is now preserved, although much machinery has been removed and a small barn which stood nearby has been demolished.

2 · Windmill design and construction

POST MILLS

The earliest form of windmill was the Post Mill. Fig. 1 shows the mill at Chillenden, Kent, which is an early design of post mill. The entire square body of the mill turns to face the wind on a very large *centre-post*. The machinery is enclosed in the body or *buck*, and the sails or sweeps drive this through the *windshaft*, which is the name for the axle which supports them. A strong ladder at the rear gives access to the interior; and projecting through the ladder is the *tailpole*, by means of which the miller can turn the mill to face the wind. A lever known as the *talthur* can be seen on the side of the tailpole; when this is pressed down it raises the foot of the ladder, which is hinged at the top, off the ground ready for turning the mill. The talthur is held in the depressed position by means of a loose pin which fits into a hole in the tailpole. The ladder is very important, for it not only provides access to the interior, but also steadies the mill and prevents the body yawing about during heavy gusts.

Fig. 2 shows a sectional model of an early post mill. If the reader refers to both illustrations 1 and 2 he will be able to follow the sequence of the construction. From the bottom upwards there are four brick piers upon which the *trestle* (see Fig. 3) is supported. The trestle consists of two large cross-timbers supported at each end by the piers. These are known as the *crosstrees*. The crosstrees cross over one another at the centre and the main post rises above them. The post has four *horns* (Fig. 4) which pass round the crosstrees and steady it at this point. Four *quarterbars* rise from the ends of the crosstrees and meet at the post just below the mill floor (Fig. 5). These quarterbars actually take the full weight of the mill, and no weight is taken on the centre of the crosstrees. There is, in fact, a small clearance at this point, and the weight passes from the post down the quarterbars and directly through the crosstree ends on to the *piers*. The trestle is very strong, and the crosstrees and quarterbars are usually of oak and about one foot thick. The post is very substantial at the horns and is often as much as three feet six inches thick; four feet has been mentioned, but the author has never measured one quite so large.

The post passes up through the mill floor (Fig. 6) and sockets into the *crowntree* in the ceiling. The crowntree passes from side to side of the mill, and has a socket in the centre into which the *pintle* (Fig. 7) of the post fits. The whole mill body turns on this point. The socket and pintle are normally of wood, but sometimes these wear and are replaced with iron. This is known as the *samson head*. The crowntree is usually of oak and very large, being about two feet thick. The crowntree at Clayton post mill, Sussex, is built up of four pieces of pitch

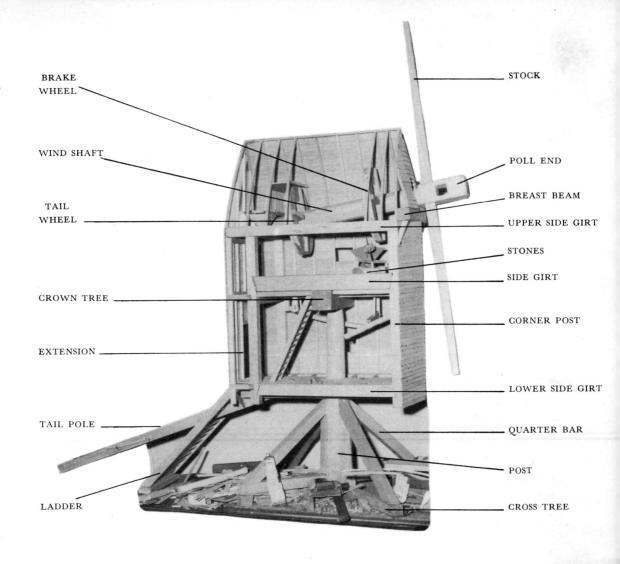

BRAKE
WHEEL

WIND SHAFT

TAIL
WHEEL

CROWN TREE

EXTENSION

TAIL POLE

LADDER

STOCK

POLL END

BREAST BEAM

UPPER SIDE GIRT

STONES

SIDE GIRT

CORNER POST

LOWER SIDE GIRT

QUARTER BAR

POST

CROSS TREE

2. A POST MILL shown in section. The Model is not complete.

pine clamped and pegged together, and measures twenty-seven inches across. The post of this mill is also built up out of four pieces of pitch pine, a very unusual arrangement.

Where the post passes through the floor there is a steady bearing called the *collar*. The two *sheer beams* pass from front to rear on either side of the post, two smaller timbers join these fore and aft of the post forming a square collar around the post (Fig. 5). On either end of the crowntree are the *side girts*. Each takes half the weight of the mill and joins on to a *corner post* at each end. Above and below the side girts are the upper and lower side girts or side rails. The corner posts are joined front and rear by minor timbers, and it will be seen that the whole construction is very strong.

3. SMARDEN, KENT Post mill trestle. Blown down early 1950s.

4. ASH, KENT Horns of main post. These remains have now gone. Blown down in 1953.

5. LOWFIELD HEATH, SURREY The post entering the mill. The floor of the body can be seen with the sheer beams and collar. Under restoration by E. Hole and Son.

6. OUTWOOD, SURREY The post inside the mill body. This is the spout floor. The crowntree passes across the top of the post. This mill is still in working order.

7. ASH, KENT The pintle of the post entering the crowntree.

8. ASH, KENT The joint between the crowntree (bottom) and the side girt.

9. NUTLEY, SUSSEX The brake wheel, windshaft and stone-nut. The brake is also visible around the rim of the brake wheel.

10. STANTON, SUFFOLK The bin floor with the top of the brake wheel and brake. The sack hoist may be observed in the roof.

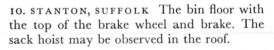

Across the top of the front (or breast) two corner posts rests the *breast beam*; this has the *chair block* at its centre which supports the neck bearing of the *windshaft*. Between the upper side girts at the rear is fitted the *tail beam*, which supports the tail bearing of the windshaft.

The sheer beams are very important, for apart from forming the collar they support the beam which in turn supports the ladder at the rear; the tailpole is fitted to them also at the rear, and in the breast they support the upright post known in East Anglia as the *prick post*. The prick post supports the centre of the breast beam.

The windshaft on early mills was wooden and octagonal or round, and tapered from breast to tail. The *tail bearing* was a simple iron pin with iron fins like a bomb. These fins fitted into deep mortices in the tail of the shaft, leaving just the pin sticking out. The result looked very similar to the pintle in Fig. 7. The *neck* consisted of a number of iron strips let into the wooden shaft, and held with clamps at either end. This neck turned in the *neck bearing* which was a U-shaped wooden bearing in early mills, but later mills used a large brass bearing resting in an iron base. When Hildenborough Mill, Kent, was pulled down (Fig. 78) the neck bearing was found to be of wood, although the windshaft was a later iron example. The bearing was actually made up of about half a dozen pieces of wood bolted together and standing on edge with the 'U' shape cut out for the windshaft.

In front of the neck bearing and outside the mill is the *poll end*. Here the windshaft is square and has two mortices passing through at right angles to one another. Through these pass the *stocks* which have a sail fitted on each end, and the stocks are held in the mortices in the poll end by wedges. These wooden poll ends often rotted after many years' use, and were cut off and replaced by an iron poll end with fins, which fitted inside mortices in the wooden shaft like the tail bearing. These iron poll ends had the advantage of not rotting and of having a smooth iron neck, rather than the iron strips. An original wooden poll end can be seen at Gayton Mill, Cheshire (Fig. 23) and an even more primitive one remains inside Portland Bill Mill, Dorset (Fig. 24).

On the windshaft behind the neck is fitted the *brake wheel* (Fig. 9), This wheel is the largest wheel in the mill, and may measure up to ten feet across. Around the rim are the *gear teeth* which drive the *stone-nut* which drives the stones. Around the perimeter of this wheel is the *brake* which is a ring of sections of wood which almost encircle the brake wheel. The brake is hinged at one end and has a long

11. (*Opposite*) CROSS-IN-HAND, SUSSEX A typical Sussex post mill with a tailpole fantail and the body covered with iron plates to help keep out the rain. The last windmill to work seriously in the county, it has now ceased. The framing is rather weak, and unless very considerable work is undertaken it may never work again.

heavy lever attached at the other. The lever is held up by a catch while the mill is working, but can be gently released from any floor by a rope. The brake is very powerful, and must be used with care, for sudden release might well cause damage to a sail.

Early windmills had only one pair of stones in the breast, driven from the brake wheel; later mills had a *tail wheel* (Fig. 2) which drove another pair of stones in the tail. Early mills also had only two floors; the *stone floor* (at side-girt level) and the *spout floor* (at collar level). Later mills had three floors with an extra floor at upper side-girt level to hold bins which contained a larger supply of grain for the stones than was previously possible (Fig. 10).

Balance is an important feature of all windmills, but especially of post mills. The post is placed forward of the centre line as can be seen in Fig. 1. The object is to balance the weight of the sails in the breast, and also the considerable weight of the stones, which can be one and a half to two tons when new. For the same reason the windshaft is inclined downwards at the tail. This brings the weight well back from the sails, and enables them to clear the cross-tree ends. The windmill is designed to receive thrust from the breast only. If a working windmill is tailwinded it can draw the windshaft forward out of the tail bearing, and as the sails rush round backwards the windshaft creates havoc as it smashes the bins and interior. In a bad case the windshaft may be torn clean out of the mill, or the mill may be blown over forwards.

If the brake is on and the mill not running it would take a very strong tail-wind to cause any damage, but it is an unwise miller who goes off to lunch and leaves the mill running unattended. The mill at Syleham, Suffolk (Fig. 96) ended its working days with two sails only, because of a tailwind in the late 1940s. The wind suddenly changed through 180 degrees while the mill was at work, and the sails ran round backwards, drew out the tail bearing and, tipping the windshaft upwards, two sails dashed themselves to pieces against the mill. Eventually two good sails and a stock were sorted out from the remains, and are still on the mill to this day. The post mill at Thornham Magna was once tailwinded too, and ended with a distinct forwards lean. While on this subject, it is as well to point out that a windmill can continue to work with two sails only, and that if one sail becomes damaged the opposite number can be removed and the mill remain at work.

Most post mills have the trestle enclosed in a *roundhouse* which provides storage space. Many later post mills are on very high brick piers. Friston Mill, Suffolk, (Fig. 39) is an example with high piers and a three-storey roundhouse.

This summarises the basic construction of the post mill. The stones, sails, etc. will be dealt with later in chapters to themselves. The next chapter deals with a more recent type of windmill: the smock and tower mill.

3 · Smock mills and tower mills

The *smock mill* was probably first used in this country during the 17th century, and is a rather more complicated type of mill than the post mill. The great difference between the two types is that only the *cap* with the sails and windshaft turns to face the wind. The tower stays stationary, and with it the machinery. If the reader refers to Fig. 12 he will see the smock mill at Stelling Minnis, Kent. Only the rectangular top or cap turns. This mill is fitted with an automatic device for turning the cap known as the *fantail;* this keeps the cap facing the wind at all times. Early smock mills were turned to face the wind by means of a tailpole (Fig. 70), but later ones, as at Henlow, Bedfordshire, and Shiremark, Surrey, were turned by means of an endless rope which ran over a wheel and down to the ground in a loop. The miller pulled on one side or another of the rope and revolved the wheel. The wheel was connected by reduction gearing to a pinion which meshed with the teeth of the rack, which had a considerable number of teeth, and encircled the top of the tower of the mill. It will be seen that, by winding the wheel round with the rope, the cap can gradually be moved to face the wind (Fig. 15). This device is known as *wheel-and-chain*, or *wheel-and-rope luffing gear.* Fig. 13 shows a very old type of smock mill at Hildenborough, Kent. This mill was turned by a wheel-and-chain gear, and the wheel may be seen standing up at the rear of the cap. The sails and chain on this mill were reached from a gallery or stage, which has since fallen off. Fig. 14 shows a much lower mill at Henlow, Bedfordshire, which did not require a stage.

Caps varied greatly in shape, but the basic early design is similar to the rectangular type found on Kentish mills. Fig. 13 shows the rectangular frame with the *cap sheers* running fore and aft on either side of the cap. The breast beam connects the sheers in the breast and the tail beam does the same in the rear. Between the sheers, just to the rear of the brake wheel, is the *sprattle beam* which is over the centre of the tower.

Around the top of the tower is the *curb*, and the *rack* is fitted to the outside of this. The curb is usually of wood, but can be of cast iron, and the cap rests directly on this. There are three types of curb: *dead curb*, *live curb* and *shot curb*. The dead curb has the cap skidding round on well-greased brass or iron pads. The live curb has the cap running on iron rollers (Fig. 16), and the shot curb has a great number of small rollers. The dead curb is stiffest to turn, but is really preferred by most millers as it stays where it is resting, and does not yaw about. There is, however, considerably less difficulty and greasing is required less frequently with the live curb. Some millwrights compromise by combining the two. Hildenborough was an example, for here there was a row of rollers

12. STELLING MINNIS, KENT This is a typical Kentish smock mill. Only the rectangular cap turns to face the wind and the octagonal tower remains stationary. Built in 1866, this is a late mill, but a post mill stood here before. The last windmill to work in Kent, it continues, using two sails only.

13. HILDENBOROUGH, KENT The framing of the tower and the cap frame are clearly seen. This brick base is of greater height than is usual with a mill of this age. Pulled down 1961.

15. HENLOW The wheel and chain luffing gear. The chain or rope is absent but the gearing is clearly shown.

14. HENLOW, BEDFORDSHIRE Both these mills had wheel and chain luffing gear. Compare the Midlands framing and cap shape with that of the south.

16. LEIGH, KENT The rollers upon which the caps of some mills turn. These run on the upper face of the curb. The remains of this mill collapsed during 1963.

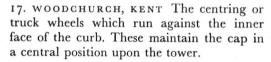

17. WOODCHURCH, KENT The centring or truck wheels which run against the inner face of the curb. These maintain the cap in a central position upon the tower.

18. STELLING MINNIS, KENT The wallower and upright shaft are seen from below. The brake wheel is shown in the top right-hand corner.

under the breast beam, the heaviest part, and two large wheels with small bearings (these could not have taken much weight) at either end of the breast beam. The rest of the curb was dead.

The cap is centred by the *truck wheels* or centring wheels (Fig. 17) which run round the inner face of the curb. The truck wheels are very important, for any slackness in the fitting of the cap would alter the position of the brake wheel and the *wallower*, which takes the place of the stone-nut on the post mill. The wallower is mounted on the upright shaft which extends from the bearing at the sprattle beam in the cap to the stone floor several floors below. If the cap was too far off centre the upright shaft would not be vertical.

The curb is supported by eight *cant* or *corner posts*, which slope upwards from the sills which rest on the top of the brick base. The slope or batter is of importance, as it gives the body or tower of the mill great rigidity. Between the cant posts is the diagonal bracing which keeps the tower rigid. The great failing of smock mills is the tendency of the tower to distort, the curb to become uneven or oval, and the cap to jam. A little decay in the joints can soon cause this trouble which, while soon cured with tie-rods and new timber where necessary, can be very tiresome. The brick base must be strong also, for this supports the sills which support the entire mill. The pressure at the sills is outwards as well as downwards, and when the sills are new they are joined at

19. The great spur wheel and stone–nut at PAKENHAM, SUFFOLK.

20. DRINKSTONE, SUFFOLK A very interesting smock mill which is reputed to have been built on a horse mill which forms the wooden base. The sails of the post mill may be just seen in the distance.

each corner, so that there is little sideways pressure on the walls of the base. If the sills are neglected and the joints weaken, the resultant outward pressure may soon crack a lightly built base. Sometimes iron bands are fitted around the base to try to remedy this fault. Bases vary from a few courses in height, as at Stelling Minnis, to several floors, as at Cranbrook, Kent. They also vary in shape from octagonal to round or to square, and many Kentish mills were built on square bases, the sills crossing the corners diagonally. Not all smock mills were octagonal, and they have varied from round to six-sided. Square smock mills have been built, but only as small drainage mills.

The upright shaft descends through the mill to the stone floor where it is supported by the *footstep bearing*. Just above the footstep bearing the *great spur wheel* is mounted (Fig. 19). Around the rim of the great spur wheel are situated the stone-nuts which drive the stones. Some later mills have the stones driven from beneath, in which case the upright shaft is longer, and the great spur wheel and stone-nuts are situated below the stone floor.

The arrangement of floors in a smock mill varies greatly, but usually there is the brick base, which serves as a spout floor, the stone floor, the bin floor and, just beneath the cap, the dust floor. The advantage which a smock mill has over the post mill is that, although it is more complicated and more inclined to leak and so decay, it can be built to a greater height. Therefore more pairs of stones

The last tower mill to work by wind in this county. This fine mill may be seen turning whenever there is a breeze.

22. HORTON CONQUEST, BEDFORDSHIRE An interesting tower mill shape. The three doors one above the other have caused the tower almost to split in two.

23. GAYTON, CHESHIRE A very old tower mill. The poll end is of wood.

can be driven from the increased wind available, and the size of sails can also be greatly increased to provide more power. Suffolk millwrights preferred post mills, and placed their mills on gigantic brick piers for the same reason.

The design of a *tower mill* is very similar to that of a smock mill, with the major exception that the wooden tower is replaced by one of brick or stone.

Although tower mills were probably invented at the same time as smock mills, they did not become popular until a rather later date. Perhaps the answer lies in cost or the difficulty of constructing the tower. The great advantage of a tower mill is that the tower is leakproof (some mills had iron plates fitted to the weather side and others were tarred to assist this) and, unless very badly constructed, do not distort. Although most mills have the doors and windows in line all the way up, this is not a good thing as cracks can develop between them. The tower at Horton Conquest (Fig. 22) had no less than three doors above one another, and the result can be seen from the illustration; the tower is splitting into two halves. To repair this type of damage can be most difficult. The strange shape of some tower mills, like Horton Conquest, is due to the tower having been built on the brick base of a smock mill. This is not likely in this case, however, as the bricks seem contemporary.

The tower of the very old mill at Gayton (Fig. 23) is very short but strong, and is built from large red stone blocks. The cap of this mill is typical of Cheshire

24. PORTLAND BILL, DORSET A primitive vertical-sided tower. This mill also had a wooden poll end which may still be seen within.

25. WENDOVER, BUCKINGHAMSHIRE An octagonal tower of heavy build. A cast-iron cross is used for attaching the sails.

and Shropshire mills. The tower at Portland Bill (Fig. 24) is typical of the South West with its vertical sides, and there are hardly any windows, so the mill must have been incredibly dark to work in. Vertical towers are not a good feature, as without a good batter there was a tendency to crack. The tower at Wendover (Fig. 25) is nearly vertical, but has not cracked, probably aided by the strong construction and careful spacing of the windows. This is a very odd mill, as the tower is octagonal, and the huge cap is, to say the least, unusual in shape.

The peak of tower mill design was reached in Lincolnshire with immensely tall brick towers with beautiful ogee-shaped caps. The tops of some of these towers were corbelled out to a slightly greater diameter just below the curb. Burgh-Le-Marsh mill (Fig. 26) exemplifies the best Lincolnshire mills.

The machinery was identical to that of a smock mill, although with the exception of Kent, where smock mills were the speciality, tower mills were built last of all, and often had very modern iron machinery. The last was built at Much Hadham in Hertfordshire in 1892.

The latest type of cap was usually round in pattern, and some late mills like East Blatchington pumping mill, Sussex, copied the Lincolnshire cap. Round caps provide a better air-flow behind the sails, but suffer from being unable to protect the cap frame from the weather. The ogee Lincolnshire type has the breast beam almost completely exposed, as are the ends of the cap sheers which

26. BURGH-LE-MARSH, LINCOLNSHIRE This five-sailed tower mill is here shown at work although it is now disused. The cap and fantail are typical of those found in Lincolnshire.

support it. No matter how well painted, these eventually decay, and the major job of replacement, sixty or more feet above the ground, has to be undertaken.

Sussex used a pepper-pot shape of cap on its tower mills, which protected the breast beam completely, as this was shorter and curved to the radius of the tower. The cap rafters, in many cases, rose up from a separate cap circle which was attached to the cap frame at suitable intervals, Nutbourne, Sussex, being a good example.

The Kentish cap undoubtedly protected the cap frame best of all, and had the added advantage of the vertical tail providing a lee which helped to protect the fantail timbers from the rain. Oddly enough many north-western mills which used a primitive form of Kentish cap had the breast beam mounted outside in the rain, Weston, Shropshire, and Willaston, Cheshire, being examples.

COMPOSITE MILLS

A *composite mill* combined the principles of both post and tower mills, for the body of a post mill was mounted in the same way as a cap on a short tower. The post and trestle were removed. There seems to be no particular advantage in this arrangement except that there was more room for storage. Very few were built, and the last remaining example was at Little Laver in Essex. Readers should not confuse this with the arrangement found in Midlands post mills, where the body of the mill had a sub-frame with rollers mounted below the spout floor. These rollers ran on a curb mounted on top of the roundhouse wall, and this curb took some weight and helped to steady the mill. The roundhouse roof turned with the mill as at Madingley (Fig. 71). A variation was the use of a small curb built up from the quarterbars, the rollers in this case being attached to the sheer beams below the body.

The *sunk post mill* was another interesting variation, which dated back to the early days of windmills. Here the trestle was simply buried in a mound of earth, presumably to provide greater stability. The snags were that not only were the sails of necessity very short, but also the trestle could not be inspected for signs of the rot which would inevitably occur. There is no example of this type remaining in England.

The last oddity is the *hollow-post mill*. Here the brake wheel drove an upright shaft which passed right through the crowntree, and down through a hole bored in the post. The upright shaft emerged beneath the crosstrees and drove the machinery in the roundhouse. The mill at Wimbledon Common, Surrey, was originally of this type, although the mill body is now mounted on a taller tower in the same way as a smock mill cap. The tiny pumping mill at Stodmarsh,

Kent (Fig. 65) is a very good example of a hollow-post mill.

Having described the various types of windmills, and their basic construction, now is the time to move on to those features common to all types: firstly the sails.

4 · The sails

The earliest known sails were cloth-covered frames and probably had a constant *weather* (the inclination of the sail frame to the direction of the wind) all the way from the heel to the tip. Later cloth or common sails had many more sail bars to support the cloth when in use, and as the heel is going slowly but the tip relatively fast, the sail was given a greater angle at the heel and a much lesser one at the tip. This is very clear in the photograph of a common sail at Drinkstone, Suffolk (Fig. 27). The same principle exactly is used on aircraft propellers. As can be seen in the photograph of Burgh-Le-Marsh mill (Fig. 26), some millwrights favoured a sharper angle at the tip, whereas others, as at Drinkstone, favoured very little. The most usual measurement was something like 28 degrees to 30 degrees at the heel, with about three degrees to five degrees at the tip. The author once constructed some sails for a one-third-size model which were 30 degrees at the heel and 0 degrees at the last sail bar. These sails were very powerful in use.

The cloth on these common sails could be furled like a curtain, and many had a type of curtain-rail at the heel to enable the cloth to be completely furled. Drinkstone (Fig. 27) and many others did not have this luxury, and a little cloth is kept on at all times. This does not matter as, since this is right at the heel, there is very little turning effect against the brake.

The *sail bars* are mortised through the *sail whip* and emerge on the other leading side. The leading side often has a *leading board* which may extend part or all the way from heel to tip. The trailing or driving side has the tips of the sail bars connected by the *hemlath*. The *whip* is bolted and clamped to the *sail stock*, which extends for about half the sail length, and stays brace the inner sail bars from the *stock*.

The sequence, Figs. 28, 29 and 30 show the erection of sails at High Salvington, Sussex. The stock is the most difficult to erect as it is very heavy, and has to be hoisted with the use of a block and tackle suspended from the poll end. The sails are hoisted in the same way and, being relatively light (four men can easily carry a small sail as shown here), present little difficulty. The greatest

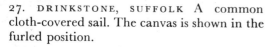

27. DRINKSTONE, SUFFOLK A common cloth-covered sail. The canvas is shown in the furled position.

28. HIGH SALVINGTON, SUSSEX E. Hole & Son fitting new sails. The sail about to be hoisted.

trouble comes when the sail has to be turned up to the top to enable the opposite sail to be fitted. Ropes are attached to the ends of the other sails, and with much heaving the sails are turned. E. Hole and Sons use winches for this job, and this is most certainly the easiest and safest method. If ropes have not been properly attached to the odd sail, there is a real risk of it turning too far and running round out of control. This should be avoided at all costs, as it could well kill someone on its way round to the lowest position.

Before erecting sails it is important to ascertain that the mill is basically sound, for this work puts considerable strain on the breast beam, much of it from the side.

Common sails are very light and very powerful, but have the disadvantage of having to be stopped in order to furl the cloth. This cloth can be spread in several positions, varying from furled to full sail. The trouble was that if the mill was turning too fast and the miller wanted to furl the cloth a little, the mill had to be stopped with the brake. This could prove impossible if a sudden strong wind sprang up, the great danger being that the mill might run out of grain and the stones run dry. This could result in a shower of sparks which could well set the mill on fire. The other risk was that the speed of the sails might become so great that the mill might be shaken to pieces by the vibration. Placed in this position, the miller would try to ride out the storm, choking the stones with too much grain to slow them, or try to turn the sails edge on to the wind, the danger

29. The sail being hoisted by a winch which is situated to the right of the photographer.

30. Only about six feet to go. The man on the sail is ready to fit the bolts.

with the latter method being that a sudden slight change of wind could cause a tailwind and blow the windshaft out of the mill. Another method of stopping the mill was to throw an abrasive, such as brick dust, on to the rim of the brake wheel. This method was used successfully by the late Mr. D. Driver when the post mill he was working, at Keymer, Sussex, ran away on bare sail frames during a particularly high gale in 1912.

A partial solution was invented in 1772 by the Scot, Andrew Meikle, this being a type of sail known as the *spring sail*. The spaces between the sail bars are larger, and in each bay two or three shutters are positioned. These shutters are hinged at either end and have a small crank fitted at one hinge. This crank is attached to the shutter bar which connects all the shutters of the sail together. If the sail is double shuttered as in Fig. 31 there are two rows of shutters, one on the leading side and one on the trailing side. The shutter bars are attached at the heel to a spring, the tension of which can be adjusted by a lever from the sail tip. The method of adjusting the spring tension varies from a simple lever to a rack and pinion, as at Ash and other Kentish mills. The springs vary also and may be full, or quarter elliptical, or even flat as at Drinkstone (Fig. 31). The most usual is the full elliptical as at Outwood (Fig. 32).

The wind presses against the shutters and tries to force them open against the spring. Thus it will be seen that the shutters partly open in a strong gust, allowing the wind to pass through. In heavy winds the miller only sets a light tension on the springs, the shutters being partly open most of the time.

31

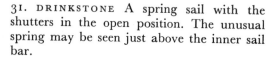

31. DRINKSTONE A spring sail with the shutters in the open position. The unusual spring may be seen just above the inner sail bar.

32. OUTWOOD, SURREY The shutters in the working position. The springs are of the more usual design.

The shutters are usually of wood, but have been made of metal in a few cases. A shuttered sail is heavy, and one method of reducing this weight was that of constructing a light wooden or metal frame for each shutter and covering this with thickly painted canvas. The shutters at Chillenden (Fig. 1) have been of this type, but the canvas has been removed to reduce wind-resistance. The whole operation of a shuttered sail is very similar to that of a Venetian blind. Spring sails do have the disadvantage of having to be adjusted individually, and the mill having to be stopped for each sail adjustment. Sir William Cubitt invented in 1807 the *patent self-reefing sail*, which was the first type of sail to be reefed from within the mill.

The construction of the patent sail was quite simple, for the shutter bars of the spring sail were connected together by the *spider*, which was positioned just in front of the poll end. The spider was connected to each shutter bar by cranks, known as triangles. The spider was connected to the interior of the mill by the *striking rod* which passed right through the windshaft and emerged at the rear of the tail bearing. By pushing the striking rod in and out, the spider operates the triangles and so opens and closes the shutters.

Self-regulation was provided by two methods. The simplest is the *rocking lever* which protrudes from the rear of the cap on most Lincolnshire and Kentish mills. A chain hangs from the end of this lever and descends to the ground, and

33. WINGHAM, KENT The striking gear, the spider and triangles are clearly shown. A clamp passes either side of the poll end to strengthen the inner stock.

34. NORTH LEVERTON, NOTTINGHAM-SHIRE Patent sails in use; the shutters are partly open in a stiff breeze. The sails are mounted on a cross.

back up to the fantail where it passes over a pulley and down to the *rocking lever*. By pulling on one or other side of this chain, the rocking lever is hauled either up or down. When hauled up the lever opens the shutters, and a weight is hung on the chain to keep them open. Sometimes a catch is used to keep the rocking lever up. By hauling the rocking lever down, the shutters are closed against the wind-pressure. The wind tries to lift the lever, but by hanging various weights on the chain the miller can regulate the effort required to open the shutters and so the speed of the mill.

The other method is that usually found in post mills, and many Sussex tower and smock mills. This uses a rack mounted on the end of the striking rod which can be moved in or out by rotating a shaft with a pinion mounted on it. The other end of this pinion shaft has a wheel (often the rim has a number of radial Y-shaped forks which give the chain a better grip), around the rim of which hangs an endless chain. By pulling on the chain the striking rod can be moved in or out; weights are hung on the chain to produce the same regulating effect found with the rocking lever. Many Norfolk tower mills use this method and have a pole which extends from the cap and steadies the chain where it hangs outside.

The chain is often operated from the top of the ladder of post mills and the wheel projects out of the tail. Many Sussex mills had the Y-wheel flat on the tail, and operated the pinion shaft by means of a pair of bevel gears. Sometimes the mechanism was fully enclosed within the mill as at Woolpit, Suffolk. Cross-in-Hand, mill, Sussex, uses a short rocking lever with a box on the end in which the weights are placed. Windmill Hill, Sussex, used a very rare device known as the *sweep governor* (Fig. 35); this operated the shutters through the ordinary wheel-and-chain mechanism. The advantage of this complicated device is debatable, but its purpose was presumably to keep the mill running at an even speed. A set of patent sails with the shutters partly opened by the wind can be seen in Fig. 34. This mill at North Leverton, Nottinghamshire, has the sails fitted to a more recent device than the poll end. This is known as the *cross*, and is a large iron casting mounted on the end of the windshaft. Wendover, Buckinghamshire (Fig. 25), has a cross clearly displayed. The cross is almost universal in the North and North East and, while usually used in conjunction with an iron windshaft, is fitted to a wooden shaft in the same way as a poll end. There are not usually stocks used with this type of sail fixing, and the whip is very much strengthened and called the *sail back*, The sail back is bolted directly to the cross.

The poll end is almost universal in the South, although West Blatchington smock mill, Sussex, uses a cross with short stocks and ordinary whips mounted on them. East Blatchington tower mill also used a cross although this pumping mill was in no way a typical Sussex mill.

Some millers sought to obtain both power, from a pair of light common sails, and some self-regulation from a pair of spring sails, as at Drinkstone (Fig. 98). Others used a pair of common sails together with a pair of patent sails as at Thornham Magna, Suffolk (Fig. 84). Some patent sails had an air brake which consisted of a leading board which opened at right angles to the direction of rotation when the shutters opened. A type of sail which is very rare today, but which was often used in Yorkshire and the North, was the *roller reefing sail*, which had a number of roller blinds instead of shutters and was operated by a striking rod in the same way as a patent sail.

With the coming of patent sails which could be operated from within the mill tower, mills were built much higher, and many sails required very tall ladders to reach them, the Lincolnshire mills like Alford (Fig. 99) being typical.

The use of the cross and sail back enabled mills to be built with more than four sails. These multi-sailed mills had five, six or eight sails and were most common in Lincolnshire. A five-sailed mill has the disadvantage of having to stop work if a sail becomes damaged, whereas a six-sailed mill is able to continue with four, three or two sails. Eight-sailed mills were not common, and are reputed to have suffered from the effects of turbulence caused by so many sails; one remains at Heckington, Lincolnshire.

Although a cross was usually used, a post mill with a three-way poll end

34

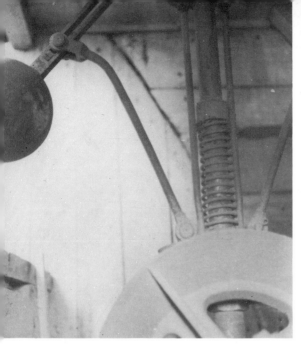

35. WINDMILL HILL, SUSSEX The sweep governor which operated through the ordinary patent sail mechanism.

37. NORTH LEVERTON The tail bearing. The striking rod with the supports for the rocking lever are just to the rear. The gear cluster for the fantail reduction drive is also shown.

36. WINGHAM The neck bearing. The bearing rests on the chair block, which in turn is supported by the breast beam. The storm hatch gives access to the poll end and sails.

38. STRACEY ARMS, NORFOLK The fantail of a typical Norfolk cap. The wheel with the Y-shaped forks is used to operate the striking gear.

stood at Ashcombe, near Lewes, Sussex. The late Mr. D. Driver from Keymer post mill, Sussex, confirmed that as far as he could remember the smock mill at South Common, Chailey (pulled down 1911), had six sails. He could not remember the type of sail fixing however. A very few annular sails have been used, but none now survive. One advantage of this type of sail and of multi-sailed mills is the smooth running produced. A post mill with four sails rolls and shakes to a surprising degree, as each sail reaches the highest point and receives the full force of the wind. When two sails are in use many millers are fearful of running the mill at full speed, for the twisting effect is very noticeable even on smock mills.

Sails have to be replaced at fairly frequent intervals. A good stock should last forty years and a good pair of sail frames fifty or even more. The stock takes a considerable strain and usually weakens at the poll end where water often lies. Bolt-holes can trap water too, and bolts should be avoided and clamps used where possible. When carrying out sail repairs it is an utterly false economy to settle for cheaper timber. A good pitch pine stock can cost £80 today, but will last a very long time. Oak stocks are very long-lasting, but are also very heavy. An Oregon pine stock may be cheaper in the short term, but if it snaps and wrecks a good sail or comes off backwards on to the mill roof, it can prove to be very expensive in the end. The post mill at Cross-in-Hand (Fig. 77) has probably stopped for good due to a stock snapping in this manner.

The sails of four-sailed mills are usually left in the form of a St. Andrew's Cross as this equalises the strain on the stocks. The greatest strain is experienced by a stock resting in a horizontal position. The vertical position is the least straining. Mills like Stone Cross, Sussex, which operated with two sails up to ceasing work about 1936, and whose miller always left them in the vertical position, often still retain the original sails to this day—Stone Cross in spite of having the extra resistance of a full set of shutters and being out of work for more than thirty years.

Mills which started life with common sails often have rather thin stocks (they are only about nine inches thick at High Salvington) and if, like Drinkstone, they have had a pair of heavier spring sails added, these are often left in the vertical position to reduce the strain on the stock (Fig. 100).

Some old windshafts were bored from end to end in order to fit patent sails. This was done both on wood and iron shafts by setting up a drill at the tail bearing. The mill was then set to work while the drill was slowly fed in by the millwrights.

5 · The fantail

The *fantail* is an ingenious device invented in the mid-1700s, its purpose being to keep the sails facing the wind at all times quite automatically. The most simple type is that seen in Fig. 38 where the cap sheers are extended at the rear and a small fan-stage built. From the fan-stage two *fly posts* rise. These are braced to the cap, and the whole construction is very sturdy. At the top of each fly post is a bearing which supports the end of a spindle, known as the *fan* or *fly spindle*. In the centre of the spindle is a hub known as the *fan star* which has six or eight *fan stocks* bolted to it. Each fan stock carries a fan blade.

The fan is connected by reduction gearing to a rack on the curb in the same way as the wheel and chain luffing gear. This gearing is very low and the fan may be revolved by using only finger pressure. A few turns of the fan will produce a barely discernable movement of the cap.

The fan is mounted at right angles to the sails, and while these face the wind, is edge on to the air flow. A slight change in the direction of the wind will strike the fan from the side, thereby causing it to rotate until the cap has once again been turned to face squarely into the wind and the fan is becalmed.

Most of the Lincolnshire type mills have an arrangement where the bevel gear on the fan spindle drives a long diagonal shaft which passes into the cap and operates a cluster of reduction gears inside, the final drive to the rack being via a pinion. The mills of the south use a different approach, for the greater part of the reduction is usually created by several pairs of gears out on the fan-stage. When the drive eventually enters the cap it is most commonly taken to the rack by means of a large worm. Some Kentish mills like Sandwich and Charing have enormous wooden worms which mesh with a rack of coarse wooden teeth with a pitch of about three inches.

Fan stages vary greatly in design from one area to another, those of Kent (Fig. 12) being small and high, while those of the adjacent county, Sussex are huge and spacious, the fans often having only five blades, as may still be seen at Polegate. The Lincolnshire fantails are the most alarming, for there is hardly any fanstage and when the miller wishes to lubricate the fan bearings, he is obliged to climb up a series of wooden blocks screwed to the high, sloping fly posts. A hand rail is provided, but the ground seems a very long way below. Sometimes a mill is equipped with a narrow gallery which extends from the fan-stage round the side of the cap to the sails. This greatly eases the painting of the cap, which can be carried out safely without the use of a cradle.

A safety device which is fitted to most mills is the provision whereby the fan can be disconnected and a hand crank used to turn the mill in the event of the fan becoming damaged. The most common method of disengaging the fan is the use of a square sleeve to connect the two halves of one of the main shafts. The miller simply removes a small bolt and slides the sleeve out of engagement. Many Kentish mills had a Y wheel and rope mounted on the side of the cap, the sole purpose of which was to provide emergency hand turning.

Should the wind suddenly change through 180 degrees in a thunderstorm, and strike the mill exactly from behind, the fan will not turn. The miller must then quickly crank the cap round a few degrees before reconnecting the fan, which will then turn the cap to face the wind. Failure to do this could result in a tailwind. Fantails are normally completely reliable and will always keep the sails facing the wind.

Post mills are often fitted with fantails, this being almost universal in Suffolk. The simplest method is that seen in Fig. 40 where the fantail is mounted on the tailpole, and the fan drives the two wheels through reduction gearing. As the wheels slowly revolve they move the fan carriage round a level track, thus turning the mill. The ladder has a pair of wheels at its foot. The wheels of the ladder at Cross-in-Hand run on an iron track. The tailpole has to be very strong with this type of fantail, and before that at Cross-in-Hand was renewed it whipped so much that the ladder wheels could be seen moving backwards and forwards while the mill was at work. In spite of adequate lubrication of the pintle, the mill also tended to jerk round with loud cracks as the fantail flexed the tailpole before moving the mill. After replacement the mill seemed far steadier while running, and the movement of the ladder wheels was greatly reduced. Only two examples of this type of fantail remain: that of Cross-in-Hand and the rebuilt one of Argos Hill nearby. Some Norfolk post mills used these fantails, but they were traditional in Sussex.

The *ladder fantail* (Fig. 39) was used in Suffolk a great deal. This arrangement spread both north and south of Suffolk, but is believed not to have occurred south of the Thames. The author has, however, seen an engraving of a Sussex post mill, where the fan looked suspiciously as if it was mounted on the ladder. An engraving cannot be relied upon as proof as it is not necessarily of great accuracy, although this one was so in other respects.

Ladder fantails usually have the fly posts inclined outwards at the top in order to allow the fan to catch the greatest wind without interference from the mill body, and also to push inwards at the foot. A little sideways skidding can occur if the mill rocks much on its post, and this outward leaning helps to

39. (*Opposite*) FRISTON, SUFFOLK The finest post mill in the county and possibly the finest in England. The size may be compared with the man walking under the breast. The fantail is typical of that favoured in Suffolk. The photograph was taken shortly before the mill stopped work for good.

40. CROSS-IN-HAND, SUSSEX A tailpole fantail being refitted to a new tailpole after the mill had been struck by lightning. A pair of new sail frames and a stock are in the foreground.

41. ICKLESHAM, SUSSEX The fantail mounted on the roof. The drive descends down the outside of the mill to tram wheels at the foot of the ladder.

counteract this. The drive to the tram wheels is the same as with a tailpole fantail. Some ladder fantails were very tall as at Friston (Fig. 39), where the height can be judged by the man walking under the breast of the mill. The drive was usually taken by both tram wheels, but at Woolpit, Suffolk, the ladder fantail drove to only one of the fine wooden wheels. The track at Woolpit was of large stones whereas some, like Syleham, were of gravel. Others were simply of large wooden blocks let into the ground.

The ladder had to be very strong when a ladder fantail was used, and to aid this much bracing was employed. Many mills had the fantail added at a later date, in which case the tailpole was usually cut off at the ladder, where it continued to take the lateral strain. Sometimes the tailpole continued through the slot in the ladder, and was attached to a crosspiece between the fly posts.

A more unusual type of post mill fantail was the *roof-mounted fantail*. This is found over a wide area and has been known in many of the more advanced windmilling counties. A fantail of this type remains at Icklesham, Sussex (Fig. 41), and the remains of another may be seen near by at Winchelsea. The fan drives a train of reduction gears in the normal way, the drive passing right down the tail of the mill, down the ladder and driving the tram wheels at the foot. The drive has also been known to pass inside the mill and turn the body by means of a worm and rack on the post. The fan could be smaller with this type of fantail, as it was more exposed to the wind than the other types.

The fan usually spends its time making only a few turns either way, and even when the wind changes it does not run continually, but tends to run in short bursts, the mill following the wind round until it settles in the new position. The behaviour of the fan in a gale is rather more dramatic, and it frequently careers round first one way and then the other. Many millers will mention this, and Mr. Dallaway from the smock mill at Punnets Town, Sussex, remembers the fan vibrating the whole cap during the gales. This was later made worse by the replacement of a broken fan blade with a modified post mill blade. This was heavier than the others, resulting in a very out-of-balance fan which used to rock the whole fan stage up and down when running fast.

The fantail is a most fascinating part of a windmill, and it is a most interesting exercise to compare the construction of those that remain. Fortunately a number remain in use, and continue to turn mills-both working and disused. Frequent greasing is very necessary, failure to do this resulting in unnecessary strain and wear. A dead curb will soon seize up if left ungreased, although the fan may still turn the cap to face the strongest gales for a few more months. Even when greased, dead curbs tend to be stiff, and the cap moves round with loud cracks. To experience a gale in a smock mill with a dead curb is quite interesting, for, with the wind roaring in the sails, beating on the cap and vibrating the tower, the fan flies round, moving the cap with great cracks which can be heard throughout the mill. On such days only the bravest of enthusiasts ventures on to the fan stage.

In light winds the fan stage or cap gallery can be a most charming place, for while the fan can be felt turning the cap and the sails cause a little movement, the sensation is in no way alarming. To stand on the cap gallery fifty or sixty feet above the ground watching the great white sails turning against the evening countryside is an entertainment no reader who has the chance should miss.

6 · The stones

A factor common to all corn windmills is the use of millstones for grinding the grain into flour. The mechanism of these will now be discussed.

The reader may remember that the top floor of a later post mill houses the bins for storing grain prior to grinding. Earlier post mills had two pairs of stones, one in the breast and the other in the tail, these being driven from above by the *stone-nut*, which meshes with the brake or tail wheel.

From the bin above, a chute, which may be of wood or a simple cloth sleeve,

42. BURGH-LE-MARSH, LINCOLNSHIRE
The stones. The hopper, horse and shoe. The
wooden spring which holds the shoe in con-
tact with the square on the quant is also
clearly visible.

43. KEYMER, SUSSEX The stones with the
vat removed. The brakewheel, stone-nut and
quant drive the runner stone.

feeds grain into the *hopper* which is mounted just above the stones on the stone-
floor. Originally these hoppers were larger and no bin-floor was used. The
miller had to keep filling them with grain continually as they only held about
two sacks, and the introduction of a bin-floor with larger bins holding ten sacks
or so was a great relief.

From the hopper the grain flows through an opening controlled by a small
slide and on to an inclined wooden trough called the *shoe*. The shoe and hopper
are mounted in a frame known as the *horse* which is itself attached to the wooden
stone casing called the *vat* or *tun*. The angle of the shoe can be controlled by a
cord which passes over a pulley and down to a knob known as the *twist peg* on
the spout-floor below. By operating the twist peg the miller can control the rate
of flow of grain to the stones. The shoe is vibrated by a square on the *quant*,
which is the vertical spindle carrying the stone-nut. This vibration causes the
grain to flow down the shoe and into the stones. The illustration (Fig. 44)
shows this very well, although Fig. 42 is the more usual type with a small shoe,
and the horse mounted entirely on the stone vat. An arrangement known as
the *bell alarm* is often found, whereby a small bell rings if the hopper becomes
empty. The most common way of operating this bell is that of having a wide
leather strap nailed inside the hopper. A string is attached from the free end
to a small bell which is suspended over a piece of moving machinery. The weight

44. DRINKSTONE, SUFFOLK The tail stones. The hopper and shoe are supported by a horse which is partly attached to the mill wall. The shoe is particularly large. This design is uncommon.

of the grain normally holds the strap firmly in the bottom of the hopper, but if this runs out the weight of the bell raises the strap which in turn allows the bell to touch the machinery. The ringing warns the miller who either stops the mill, or quickly takes some grain up to the bins before the stones run dry.

There are two millstones, one stationary and imbedded in the floor and the other *runner stone* which revolves above it. The runner stone has a round hole in the centre known as the *eye*. The grain pours from the shoe down through the eye, and is ground as it passes between the surfaces of the stones. Flour emerges at the outer edge where it is swept round inside the vat by a small paddle until it drops down a spout to the floor below.

In England two basic types of stone are found—the *peak*, which is a one-piece stone mined in the Peak District, and the *French burr*, which is built up from a number of wedge-shaped pieces of hard quartz. Peak stones are rather coarse in texture and are used for cattle-food. The French burr is smooth and very hard and was used for flour-production. The wedge-shaped pieces of stone were arranged in a special manner, and were both cemented and clamped together with iron bands around the circumference. The non-working surface was usually faced with plaster of Paris. Another imported stone, which is rarely found today, in the *Blue* or *Cullen stone*. This was a one-piece stone, with a rather smoother texture than the peak, which was imported from Germany. A stone often found in the North West was the *red stone* which had a texture like very hard sandstone. This was also a one-piece stone, mined locally. A pair of these stones may be seen at Gayton, Cheshire (Fig. 23), where the tower appears to be built of the same type of stone.

The grinding surface of the stones has a number of tangential grooves cut in it which convey the flour to the periphery. On each *land* between the furrows are cut a good number of fine grooves which actually carry out the grinding. The surface of the stone gradually wears, and the runner stone has to be lifted up for dressing. This is done by removing the vat and either winding a rope around the windshaft (where deep grooves are often worn), and using wind power, or using a block and tackle. This latter method has to be used in tower and smock mills due to the windshaft being out of reach. The stone is carefully laid on its back ready for the miller, or in the past the stone-dresser, to commence the skilled job of stone-dressing.

The furrows are deepened and the *stitching*, as the fine dressing is called, re-cut by means of a small handle which holds a very hard chisel-shaped *mill bill*. When grinding cattle-feed the *thrift*, as the handle is called, holds a pointed mill bill called a *pick*. This chips out small indentations in the lands. The surface must be level, and this is tested by drawing a wooden staff across the surface. The face of the staff is coated with red oxide which leaves a red mark on

45. WILLESBOROUGH, KENT The bed stone with dressing tools. The staff stands at the rear. A mill bill is in a thrift on the far side. A spare thrift and a few mill bills and picks lie on the stone. The jack staff is mounted on the stone spindle. The mace, a wedge and a crowbar complete the dressing tools.

46. SAXTEAD GREEN, SUFFOLK Two pairs of stones are driven side by side in the breast.

the high spots. The miller then chips these high spots away with the mill bill or rubs minor ones down with an old piece of burr stone, called the *rubbing burr*. The runner stone must be slightly concave towards the eye to encourage all the grain to enter the gap. A small *eye staff* is used to test the level of the concave area. The staffs are tested occasionally against a proof staff which is very carefully preserved. One at Meopham, Kent, was of slate although others were of iron. Any high spots or signs of warping on the staff were remedied with great precision. The dressing tools may be seen in Fig. 45. The stitching is done after the stone has been levelled.

The runner stone is supported from below by the *stone spindle* which passes up through a bearing in the bed stone. The stone spindle must be vertical or the runner stone will not run true, and to test this a wood or iron bar, called the *jack staff*, fits to the upper end which projects above the bed stone. A feather is fitted to the end of the jack staff and, with an assistant turning the spindle from beneath, the miller is able to observe whether or not this touches the bed stone equally all the way round. If it does not then the bearing at the foot is adjusted by wedges or screws until the spindle is vertical.

A leather washer prevents dust entering the bearing in the bed stone. Above the washer the stone spindle rises as a tapering square, and the *mace* fits on to the square in such a way that the tip of the spindle projects through with a

47. SHIPLEY, SUSSEX The governors which control the gap between the stones according to the speed of the mill. Some late mills used one set to control all the stones.

48. CROSS-IN-HAND, SUSSEX A mortised iron great spur wheel drives a stone-nut underdrift in the breast. A cast-iron bridge tree with screw for disengaging the stone-nut is to the right.

rounded tip. The runner stone has a bar across the eye which has a recess in the centre, and the recess rests on the *cock head* (the rounded tip) of the stone spindle. The bridge in the eye of the runner stone engages with two recesses in the mace. Thus the runner stone is balanced on the cock head, but turns with the stone spindle.

The quant which carries the stone-nut has a large fork formed in the end which engages with two more recesses in the mace. The upper end fits into a bearing in a *sprattle beam* just to the rear of the brake wheel. There is some slackness in the fitting of the quant at the mace, which enables the sprattle bearing to be released and the stone-nut moved out of engagement with the brake wheel. This allows either pair of stones to be used as the miller wishes.

Later post mills often had two pairs of stones side by side in the breast, in which case the drive was through a wallower, upright shaft and great spur wheel as in Fig. 46. This method was much favoured in Suffolk where a third pair was sometimes driven in the usual way from the tailwheel. Some large post mills had four pairs of stones, with two pairs driven by both the brake and tail wheels.

The stone spindle rests in a footstep bearing which is situated on the *bridge tree* in the ceiling of the spout floor. The bridge tree is pivoted at one end and is supported at the other by a lever known as the *brayer*. The free end of the brayer is supported by the end of a long iron lever called the *steelyard*. The steelyard

46

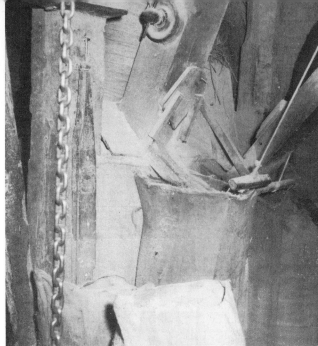

49. STANTON, SUFFOLK The stones, under-drift in the breast. A wooden upright shaft is used.

50. NORTH LEVERTON Barley meal pouring into a sack.

passes over a fulcrum and, being of considerable length, has very great leverage. A movement of one inch at the steelyard end produces nearly imperceptible movement of the stone spindle in a vertical direction. The miller can adjust the gap between the stones by means of a screw at the junction of the steelyard and the brayer. After starting the mill the miller stands here feeling the flour as it emerges from the spout with his left hand and adjusting the screw with the other until the best setting is achieved.

In early days the miller had to continue adjusting the stones all day, for as the speed increases the runner stone tends to rise a few thousandths of an inch. Small as this measurement is, it is none the less able to alter the quality of flour, and the miller had to lower the stones a fraction. A gusty wind would require the miller to alter the adjustment almost incessantly. To avoid this trouble a governor (Fig. 47) was installed. This was usually belt-driven from the stone spindle, and as the speed of the mill increased the weights flew out causing a collar on the spindle to rise. A fork on the end of the steelyard engaged with this collar, and rose and fell as the speed changed. As the brayer and bridge tree were connected to the other end of the steelyard, it will be seen that the stone spindle also rose and fell quite automatically.

Many late windmills drove the stones from below—a method known as *underdrift* or underdriven. The arrangement was less common in post mills than in smock and tower mills, but Fig. 48 shows such an example at Cross-in-

47

Hand, Sussex. The only difference was that the stone-nut was mounted on a square on the stone spindle and was raised out of engagement with the great spur wheel by means of a ring which pushed the stone-nut up the spindle and out of mesh. The screw operation of this can be seen together with an iron bridge tree in Fig. 48. In place of the quant a small iron forging called the *damsel* was used to vibrate the shoe. This can just be seen projecting above the right-hand stone vat in Fig. 49. The damsel was so called because, with its three or four arms striking the shoe, it made more noise than any other part of the windmill's mechanism.

An odd arrangement found in some Midlands post mills is the *hurst*. The hurst is a stout framework which supports two pairs of stones, underdrift, side by side in the breast on the spout floor. The bed stone is only about four feet above the floor, and there is some difficulty in providing room for a reasonably large bin to receive the flour. The whole mechanism is very cramped and it is hard to see the advantage, if any, over the other types of stone situation.

7 · Other machinery

SACK HOISTS

To raise the grain to the bin floor the miller used the *sack hoist*. The simplest form of post mill sack hoist was that with a long roller mounted in the peak of the roof on the bin floor. A pulley was mounted on one end round which a slack belt or chain passed. The belt also passed round a pulley on the windshaft and, by using a jockey pulley or a system of levers which raised the pulley end of the sack hoist roller, the belt was tightened, thus revolving the roller. The other end of the roller had a chain attached which passed right down through the mill to the ground. When the hoist was used the miller usually tipped the sacks on the bin floor while the mill boy slipped a noose at the end of the chain round the neck of the sack. The control rope passed down to the ground too, and a jerk would signify that the sack was ready. The miller then operated the control cord, and the sack came up through a series of trap-doors so designed that the double flaps of which they were comprised fell shut behind it.

At the bin floor the sack could be balanced on the edge of the bin by slipping the drive, and provision is always made for this. The chain was then dropped to the ground through a good-sized hole at the joint of the two halves of the trap-doors. Tall post mills suffered from the difficulty of not being able to raise sacks from the bottom floor of a multi-floored roundhouse unless the mill faced in certain directions. The roundhouse roof does not meet the post, but joins a ring which is supported from the quarterbars at about a four-foot radius from the post. Sacks can, therefore, be raised from a single-storey roundhouse always, although often having a bumpy ride, past the quarterbars on their way up.

Sack hoist drives are many and various, but even if the power take-off is through a gear, there is a friction arrangement in the system somewhere which allows the drive to be engaged without stopping the sails. Smock mills and tower mills also use very many forms of sack hoist drive, the most usual being the mounting of the sack hoist roller on the dust floor with a drum on the end engaging with a friction surface on the lower face of the wallower. These drums varied from about eighteen inches in diameter to nearly four feet as at Hilden-borough. The sack hoist drive at Silver Hill mill, Hastings, was gear-driven from just above the great spur wheel, with the drive passing upwards via a leather belt. The sack traps were usually all in perfect line right up the mill and could be heard banging shut as the sack passed through five or more floors. Some rather small mills had a hinged ladder to the bin floor, which could be moved to the vertical position and the aperture used to raise sacks, normal sack traps being used for the floors below.

BOLTERS AND WIRE MACHINES

When the flour emerges from the stones it is quite warm and is allowed to cool in wooden bins on the spout floor. The miller then bags it up and hoists it to the bin floor, where he empties the sacks into another bin. This bin originally fed the *bolter* which consisted of an inclined wooden frame covered by a woollen sleeve. The mesh of this sleeve was fine at the top and became coarse towards the bottom. The cylinder was rotated by a pulley attached to a shaft running through the centre.

The coarse flour was fed in at the top end by a shoe similar to that used at the stones. As the cylinder revolved the finest flour passed through the first mesh and was collected in a chute. The coarser flour was collected in the next, and the very coarse flour in the next, while the bran tailed out of the end. To en-courage the passage of the flour through the cloth, a number of springy wooden bars were positioned to rap against the cylinder frame, which was thus vibrated. A bolter is rarely found in a windmill today, although the barest remains of one may be seen at Shiremark, Surrey.

The bolter was superseded by the *wire machine*, which, although operating on the same principle, differed in that there was a fixed cylinder covered with various meshes of wire. The spindle drove six brushes or more, which revolved inside the cylinder and brushed the flour through as the meal descended from the shoe. A wire machine with the access covers removed may be seen in Fig. 51. The various grades of flour emerge on the floor below, from chutes which feed directly into sacks.

The *jog scry* was an older form of flour dresser, which consisted of an inclined chute with several grades of mesh forming a false floor. One end was hinged, while the other was jerked up and down by a long belt attached either to a short throw crank or to a lever which was moved up and down by a number of cams, looking similar to a very coarse gear wheel (Fig. 61). That shown in Fig. 52 was used to grade the bran as it emerged from the wire machine.

Other machines found were *smutters* (Fig. 53) which attempted to remove the black fungus which sometimes attacks wheat, and *grain cleaners* (Fig. 54), which often did not have the fan, seen here, for blowing away the dust. Argos Hill mill, shown here also, had a wind-driven grindstone, which would have been used to sharpen the miller's mill bills.

The problem of space was not great with smock and tower mills, where such machines were easily fitted in. Post mills usually had an extension (Fig. 2) to the rear of the corner posts in which these machines could be housed. Argos Hill mill has an extension upon the extension, this reaching only about two-thirds of the way up. Bolters and wire machines were nearly always mounted crossways in the tail of post mills and were rather shorter than that shown at Capenhurst. The drive to bolters and wire machines varied greatly but was usually taken from a bevel ring on the great spur wheel, or from a separate bevel gear (Fig. 55) mounted on the upright shaft in tower and smock mills. Post mills often had a skew gear mounted on a crossways shaft and driven from the brake wheel, the drive then being by long belt to a pulley in the tail.

GEAR TYPES AND VARIATIONS

Some of the greatest skill was required by the millwrights when making the wooden gears. These had to be very strong and yet perfectly round, with the teeth spaced at the correct pitch. The earliest type of brake wheel was that known as the *compass arm* type, where two spokes were mortised right through the windshaft in the form of a cross. To the ends of these spokes four or six *cants* were fitted which formed the rim. The ends were joined with tenons and bolts, while more curved pieces of timber formed a ring which crossed the joints on either face.

A large number of wooden pegs were mortised right through the rim and emerged on the far face, where they were pinned. These pegs engaged with a stone-nut of the lantern pinion type. This consisted of two solid wooden discs

51. CAPENHURST, CHESHIRE A wire machine. The brushes are in position inside the
cylinder.

52. BEXHILL, SUSSEX A jog scry used for
grading the tailings from the wire machine.

53. SHIPLEY, SUSSEX The smutter.

54. ARGOS HILL, SUSSEX The grain cleaner.
The lower pulley drove a fan to blow away
the dust.

55. BARDWELL, SUFFOLK The auxiliary drive.

56. ROLVENDEN, KENT A fine wooden stone-nut.

57. WINCHELSEA, SUSSEX A brakewheel of refined construction. The windshaft is unfortunately missing.

58. CLAYTON, SUSSEX This tailwheel is of very heavy construction and has been converted from compass arm design to that of clasp arm.

joined by a number of staves. The result was similar to a cylindrical bird-cage or an old-fashioned lantern. The quant passed through a square hole in the centre of each of the wooden discs.

The *trundle wheel* was a development of the lantern pinion in which the top disc was removed leaving the staves projecting upwards. These engaged with the brake wheel pegs in the same way as the lantern pinion.

The later wheels had proper wooden teeth around the rim which engaged with an early type of wooden bevel gear. This type of stone-nut was bound with iron and appeared very similar to a wooden cart hub (Fig. 56).

The compass arm wheels were not as easy to keep true as some later types, and surprisingly few remain to this day. Many wallowers were compass-arm, and often had the upper and lower surfaces covered in with thick planks, the result looking like a solid wheel. A pair of compass arm wheels meshing together is very rare, although a compass arm brake wheel and wallower may be seen at Weston, Shropshire. Wallower rims were frequently laminated to reduce the possibility of splitting.

Most wooden brake and tail wheels are of the *clasp arm* type where the timbers form a square through which the windshaft (Fig. 57) fits. Many later windshafts had a square formed at each wheel position for this purpose. A good number of mills have had compass arm wheels converted to clasp arm, as has the wheel at Clayton (Fig. 58). It is interesting to compare the heavy construction of Fig. 58 with the lighter construction of Fig. 57; the latter is probably the younger wheel.

Great spur wheels were also of compass or clasp arm construction. Most compass arm wheels had four spokes (Fig. 59), but some had six as at Nonnington, Kent, and others had eight as at Hildenborough in the same county. The four intermediate spokes on this last did not pass right through and only entered the upright shaft for a depth of about four inches. Compass arm great spur wheels were not uncommon in Kent, although most had four arms, more than this number presumably causing a weakness of the upright shaft.

Clasp arm great spur wheels are often seen, a very fine example being at Chislet (Fig. 60), where the entire weight of the upright shaft and stones was taken by a very substantial hurst frame. This method is rare, and it would be quite possible to dismantle the entire tower, but still leave the stones and upright shaft standing.

By far the most common constructional material to be found in windmill gears today is iron. Many old wooden gears have been replaced by cast iron, and some have had the teeth cut off and iron teeth bolted on in sections (Fig. 44). Most of the remaining stone-nuts are of cast iron. Cast-iron gears last almost indefinitely but, if meshing with another iron-toothed gear, can be extremely noisy. Pakenham Mill, Suffolk, used to have a small pair of stones driven by an iron stone-nut from an iron internal ring on the great spur wheel (Fig. 19). The noise was most striking and was made far worse by the ring being out of true. A local millwright blamed this on the wooden pattern warping before being used to make the mould.

To reduce the noise many millwrights fitted wooden teeth to iron wheels; these were known as mortised iron wheels (Fig. 61 and 48). Wood to wood, or wood to iron is an almost silent drive, and only the faintest purring is audible. Wooden teeth were usually made of apple wood, although other tough hard woods like holly and hornbeam were also used. These teeth do not wear very much, and if well lubricated with grease or linseed oil will last half a lifetime.

The shanks of the teeth usually projected through the rim and were held with two small nails driven in towards each other (Fig. 58). An older method was that of driving wedges between the shanks as in Fig. 61.

A unique arrangement may be seen in Fig. 62 where the wooden teeth of the brake wheel drove the iron wallower through two staggered rows of teeth. The purpose was probably to reduce backlash and achieve greater silence of operation.

The study of auxiliary drive gears is also most rewarding, for these are of extreme variety. Sometimes a pair of well-made wooden bevel gears may be seen as in Fig. 55, or sometimes just an iron bevel and an iron ring on the great spur wheel. The drive to the wire machine at Hildenborough was from a bevel ring on the great spur wheel, and then through a pair of wooden bevel gears and finally by belt.

Very fine wooden skew gears were to be seen in many post mills driving the auxiliary machines from the brake wheel. To eliminate the skew gear characteristic short shafts were sometimes used as at Lowfield Heath, Surrey, when it was possible to mount the pinion on the radius. The teeth of a very few post mill brake wheels were also set at a slight angle from the radius to enable the wallower to be situated off centre. This enabled a larger pair of stones to be used on one side of the breast.

Two cast-iron windshafts, both found in Sussex, had mortises right through to facilitate the use of a compass arm brake wheel. One shaft was from Clymping smock mill, and the other from Sullington Warren post mill. There seems no reasonable explanation for this practice.

59. WHITSTABLE, KENT The compass arm great spur wheel.

61. FRISTON, SUFFOLK The iron tailwheel. The teeth are held in position by wedges driven between the shanks. The gearlike cam ring drove the jog scry.

60. CHISLET, KENT A clasp arm great spur wheel. The auxiliary drive bevel gear is driven from an iron ring bolted to the lower face.

62. MADINGLEY, CAMBRIDGESHIRE The curious staggered teeth of the brake wheel and wallower are clearly shown.

8 · Marsh mills

Windmills have been used for pumping for several centuries, and while a few like that at Faversham, Kent, were used for raising water from wells, their main use was for drainage.

Wind pumping mills were used in the years from the late seventeenth century onwards, becoming very popular indeed during the eighteenth and nineteenth centuries. Wind pumps were still being built in very limited numbers into the first years of the present century.

The most usual type found is the tower mill, although there were many smock mills in the past. The brake wheel drove the wallower on a very long upright shaft which terminated in a bevel gear on the ground floor. This gear drove a shaft which passed out of the mill and into a narrow semicircular casing positioned over a culvert outside the mill. Inside the casing was a large-diameter wheel with paddles attached around the rim. The paddles fitted the culvert closely, and when the wheel was rotated by wind power the water was swept up to a higher level where it flowed into a canal. The scoop wheel could not lift water very far, but was simple in operation and was the mainstay of fen pumping for many years. Plunger pumps lifted further, but were mainly used in connection with wells, their use on marshes being rare.

The *Appold turbine* (Fig. 64) was a more recent type of marsh pump, and worked in a cylindrical casing in the culvert. The vertical shaft carried a num-

63. HORSEY, NORFOLK A well-restored drainage mill although the striking gear for the patent sails was not replaced. There is a stack of thatching reeds in the foreground.

64. STRACEY ARMS, NORFOLK An Appold turbine.

ber of vanes which caused the water to rise up and flow out of the casing at a higher point. The principle is similar to that noticed when a cup of tea is stirred too fast.

Some very strange contraptions were used as wind pumps, not the least peculiar being the use of bare smock mill frames winded with tailpoles and carrying large common sails. Stodmarsh, Kent (Fig. 65) typifies a very rare type of hollow post mill, and should be restored, as unless one seen near Acle, Norfolk, still survives, this is the sole remaining example. Many dozens must have existed in the fens, and yet this survivor is in a county where virtually no wind drainage mills existed.

Few complete drainage mills remain, and only Wicken Fen, Cambridge-shire, and Herringfleet, Suffolk, work occasionally, and then only for demon-stration purposes. The rate of deterioration seems high, for, although most have the characteristic Norfolk-shaped cap which protects the timbers well from the elements, the majority remain as empty towers. A few mills were left high and dry by the receding water and were converted to corn mills. These must have worked well, for flat areas are good for wind.

HORIZONTAL MILLS

While on the subject of unusual windmills, it might be as well to say a few words about horizontal windmills. Examples have been built at both Margate

65. STODMARSH, KENT There were once many pumping mills of this type in Norfolk. This may well be the sole survivor in the country. The drive passes down through the hollow post.

66. SAXTEAD GREEN, SUFFOLK The sails are idling in a good breeze. The whole mill body was rebuilt by the Ministry of Works to a very high standard.

and Battersea, both to the design of the same man. A tall tower was built in the manner of a lightly constructed smock mill tower. The machinery was contained in the lower floors, while the upper part had vertical louvres around the circumference. The wind blew through these and activated a large horizontal paddle-wheel, which was attached to an upright shaft. The louvres were inclined at an angle to enable the wind to impinge on only one side of the paddle-wheel no matter from which direction it was blowing. Nothing of these two mills remains. Robert Beatson, in his *Essay on the Comparative Advantages of Vertical and Horizontal Windmills* (1798), shows a horizontal windmill of his own invention. A shaft projects through the roof of a building and carries four blades. Each blade is constructed with shutters like a patent sail. The idea was that the wind would open the shutters from one side, but close those on the other. The difference in wind-resistance would cause the blades to rotate the shaft, and as each open blade came round the wind would close the shutters as it struck it from the opposite side. It would be most interesting to know whether any mills of this type were constructed.

9 · Windmills in general

Most windmills are situated on hills or in other wind-catching positions. Flat areas are just as good for wind, and those mills in Lincolnshire or the fens worked as well as any. There is no explanation for the curious Shropshire practice of placing the tower mills of the area on the lower slopes of quite high hills. The mills must have worked well from some directions but have been completely sheltered from others. Vennington is an example, and there is even one empty tower on the lower slopes of the famous Wrekin, which rises a matter of hundreds of feet above. This practice was not universal in the area, and some mills like Rowton and Homer may be seen for miles.

An unobstructed position with little turbulence caused by trees or buildings is very important, and several mills were jacked up bodily on to higher brick piers or bases. Tower mills were raised vertically from the original curb level (Fig. 87). The reasons for raising mills varied, but the most usual was that of growing trees or encroaching houses. At Drinkstone, Suffolk, the post mill stands on a mound while the smock mill is situated on the flatter ground a short distance away. It is interesting to note that, although the post mill mound reached to a fair height, this did not shelter the smock mill from the wind, and it worked perfectly when facing in this (northerly) direction. When the post mill started, however, it created turbulence, and the smock mill would not run satisfactorily. The miller of Sandwich smock mill, Kent, claimed that the turbulence from a shed built a few hundred yards to the south by the railway had an effect on his mill. When the iron brake wheel spokes broke one day, proceedings were taken against the railway company. To prove the miller's point, a bonfire was started to windward of the shed and the smoke was duly observed eddying across to the mill. The miller's wind was originally protected, but as time elapsed it became evident that the miller's only real method of dealing with obstruction was either to raise the mill or move it to a better position.

Many wooden mills were moved in the past, either in one piece or in parts. Post mills were frequently moved about in this manner, when they were dragged on rollers or specially constructed sledges by as many as three dozen horses or oxen. Cross-in-Hand (Fig. 11) has been moved twice. Smock mills were either sawn down each corner post and moved in sections, dismantled, or moved with the body on timber carts and the cap and machinery separately. This was not possible with tower mills, where the tower had to remain while the machinery and cap were fitted to a new tower at the new destination. Probably the most famous removal was that of the post mill at Clayton, Sussex, from a position at Brighton to the site it now occupies. This was achieved with a great number of oxen, and an old etching shows the sails off and the trestle resting on a wooden frame to which six rows of oxen are attached by traces. The mill body

67. ROLVENDEN, KENT Restored as a memorial. Much work was carried out in the mid 1950s. A very good restoration for the period. The well-known millright and enthusiast V. G. Pargeter is seen standing on the poll end.

68. FRAMSDEN, SUFFOLK Shown here before restoration was started in the mid 1960s.

is braced to the frame by long timbers which extend down from the upper parts of the mill body. When the size and weight of a windmill is considered, a removal would be an outstanding achievement even today with modern cranes and equipment.

PAIRS OF WINDMILLS

Windmills were often found in pairs, which usually resulted from trade increasing to such a degree that the first mill could no longer cope. For this reason the mill which was added was often larger and much more modern than the original one.

This explains why most pairs are of different types and certainly different dates. The post mill at Clayton was added to a very old post mill which was itself demolished and replaced by the tower mill during the later nineteenth century. The single-storey roundhouse of this mill may still be seen at the base of the tower mill. The mill at Cross-in-Hand was also added to an old post mill, the roundhouse of which may still be seen. Examples of smock and post mills together were to be seen at Fornham and Drinkstone, Suffolk, and Outwood, Surrey. The remains of one such pair still exist in the grass at Great Hormead in Hertfordshire.

To see a pair of windmills at work must have been impressive, and people who have experienced this are few today, although an old Sussex carter could well remember such a sight at Clayton. How disappointing it is that nowhere can a pair of English mills be seen both with sails. To fit these at Clayton tower mill would be very easy, despite the absence of the brake wheel. Drinkstone smock mill is worthy of restoration too, for although the windshaft and machinery are missing, the finding of a suitable windshaft would not be difficult. This mill is very old, and is mounted on a wooden base which is reputed to have been a horse mill. Due to slackness in the joints between the cant posts and upright posts of the base the whole mill swayed slightly while at work.

OLDEST REMAINING MILLS

Almost certainly the oldest remaining mill is that at Bourn in Cambridgeshire. This old post mill is small in size and is original even to the open trestle and pent roof. This mill has, fortunately, been preserved since the 1930s when it was proved to have been in existence in 1636. The oldest dated mill is that at Pitstone Green in Buckinghamshire, which has the date 1627 inscribed within. There seems little doubt, however, that Bourn mill in fact antedates this structure. The oldest dated working post mill is that at Outwood, Surrey, which worked from 1665 to the early 1960s when the miller, Mr. Jupp, died. For-

69. OUTWOOD, SURREY A rare sight today, a pair of windmills with one at work. The smock mill was blown down in 1961, but the post mill, which was built in 1665, remains in use.

tunately the new owner continues to run the mill for demonstration purposes and the interior may be viewed for a small charge. One of the oldest windmills still working is that at Drinkstone, Suffolk, which has vertical side girts which engage with joints on the upper and lower side girts. This is the oldest form of construction, and is similar to Bourn in this respect.

Some later post mills also used this form of construction which cannot be relied upon as absolute proof of extreme age. Drinkstone mill has had a varied life, having started with a square body only nine feet across (measurements given by Mr. Clover, the miller) and been extended both in the breast and tail. The result is that the post is central and the mill rather out of balance, causing it to be a little 'headsick' (lower at the breast); a 'tailsick' windmill leans backwards. The central position of the post also causes more swaying while at work than is experienced in most post mills.

Some areas were less inclined to change and accept new ideas than others, with the result that their windmills appeared older than was in fact the case. Some older mills have clockwise sails, although Burgh-Le-Marsh, Lincolnshire, is a clockwise five-sailed mill—a feature quite out of keeping with its other Lincolnshire characteristics. The great majority by far had anti-clockwise sails, and second-hand stones have to have the dressing cut in the reverse direction when used in a clockwise mill like Drinkstone. Unless the stone-nut is forward of the brake and tail wheels (and they seldom were) the stones turn in the same direction as the sails when viewed from above. When an upright shaft is used as was always the case in tower and smock mills, and was also found in

70. WEST WRATTING, CAMBRIDGESHIRE
An old smock mill winded by a tail pole. The
cap has been rather altered and two clockwise and two anti-clockwise sails have been
fitted.

71. MADINGLEY, CAMBRIDGESHIRE A
Midlands-type post mill moved from Hun-
tingdonshire to its present situation before
the war.

72. BARHAM, KENT The sails are idling in a
good breeze. The last few sail bays do not
contain shutters in order to reduce wind
resistance. This original and well-preserved
mill was completely burnt down during the
winter of 1969-70.

many post mills, the rotation is the opposite to that of the sails. For this reason
tower millstones could be used in a clockwise post mill without re-dressing.

LOCAL TRADITIONS AND VARIATIONS

A comparison of windmills from various areas is a rewarding practice, for
quite definite local traditions may be observed. Such features as the Suffolk
ladder fantail are very obvious, but the more subtle variations are of even
greater interest. The general rule may be said to be that the windmills of the
Eastern counties are very advanced, while those of the West become more and
more primitive in construction. Lancashire rather disproves this, for the mills
are quite modern, although distinctive.

A county by county analysis follows, but owing to the passage of time and
scarcity of windmills in some counties it should not be regarded as absolutely
complete. For reasons of accuracy only personal observations, both from old
photographs and from windmills themselves, have been included. Hearsay
cannot be relied upon as fact any more than can conjecture.

Starting with the South East the types and variations may be classed as
follows:

Kent

A land of tall smock mills with caps similar to post mill roofs. Shuttered sails and a rocking lever, when the sails were patent, were common. Square brick bases were frequent, and broad stages were very common indeed. Fan stages were of an individual type, and had fans with six or eight blades well out from the fan star. The blades were small and the fan appeared rather sparse.

Sussex

On the whole a mixed county with a special type of pepper-pot-shaped cap on the tower mills. Smock mills seemed to favour the Kentish shape of cap. Many shuttered sails were used and could be of either spring or patent type. A wide leading board was often used for about the first two-thirds of the leading edge, the tip having shutters. Instead of standing out on the front of the whip, the triangles which operated the shutter bars were frequently found on the side of the stock. The spider thus had a rather twisted look when viewed from the ground.

The post mills were large, often with the tailpole fantail already mentioned. The breasts of these mills had a very pronounced point which aided air-flow from the sails.

Surrey

So few mills remain in this county that it is not really possible to draw many conclusions. The evidence of old photographs certainly seems to indicate, however, that most of the smock mills at any rate had Kentish-type caps. A plentiful supply of water exists in this county, which encouraged the use of water mills. This, and the close presence of London, may help to explain the marked absence of windmills within the photographical past.

Somerset and the South West

Very few mills remain in these parts, but the Somerest type is the most common and was usually a vertical tower of fairly large diameter. The caps were large and based on the rectangular pattern with steeply sloping rafters. A wheel-and-chain luffing gear was usual. These mills were rather primitive in design.

Essex

Had a mixture of types of mill and did not appear to have a very distinctive type of its own. The ladder fantail was found on several post mills.

Suffolk

Here the post mills were of a very individual type, being tall and narrow, often on very tall brick-built roundhouses. The breasts were frequently curved with the boards bent round. Two pairs of stones side by side in the breast were also common, and shuttered sails were usual and were likely to be of the patent type. The timbers were often surprisingly small in section. Tower mills often seem to have had a rather tall beehive-shaped cap. Caps certainly tended to be round. Smock mills were uncommon.

Norfolk

Tower and drainage mills were common here, and had a very distinctive type of boat-shaped cap which is best described as a more streamlined Kentish design. Rack-and-pinion-operated striking gear with a pole to steady the chain was usual.

Cambridgeshire

This county contains some very interesting mills indeed, many being of a rather individual type. The common factors are few, but caps tended to be of a special pointed type, rounded in pattern with vertical boarding. The fantails had vertical fly posts which projected well above the small horizontal fan stage at the rear. The fans were large, and braces passed from the fly post tops to the top of the cap. Vertical boarding was sometimes found on the smock mills.

Lincolnshire

This county achieved the peak of tower mill design with tall towers which had attractive ogee-shaped caps. The cross was the most common method of sail attachment, while patent sails were numerous. A rocking lever was used to operate the striking gear. Fantails were most outstanding, with inclined fly posts braced to the small fan stage. A hand rail is provided along the side of most fly posts for use by the miller while maintaining the fan gears and bearings. Fans were quite large with about eight blades. Machinery was often of iron and very modern.

The North East

Again predominantly tower mills showing a strong Lincolnshire influence. Roller reefing sails were sometimes found in Yorkshire.

The Midlands

An area showing considerable influence from the surrounding counties, the tower mills of the East showing a Lincolnshire influence whereas those of the West were more primitive. The post mills were quite outstanding with many special features. Breasts were often flat, and the breast beam straight with its ends resting on a forward projection of the upper side girts. When standing by the breast beam on these mills it is often possible to see a gap of several inches or more between the beam and the weatherboarding of the breast. A curb was also often seen on the trestle or roundhouse wall. The roundhouse roof also frequently turned with the body. Hurst frames were common, with two pairs of stones side by side in the breast of the spout floor. The framing was of large timbers with less diagonal bracing than was seen in the East. A very few in the extreme South of the area had three crosstrees and six quarterbars. These post mills seem to have had a remarkable resistance to the weather, for many old photographs show the mills still standing with virtually no boarding remaining. The use of a curb and oak framing may be an explanation.

Lancashire

In this county many tower mills were built. The towers were often white with black caps of a rectangular pattern. Shuttered sails were common, as was a cross to mount them. Fantails were high with sloping fly posts. None of the primitive post mills remains.

Remainder of the West

Primitive tower mills were of the usual type with wheel-and-chain luffing gear and common sails. Windmills are very scarce in the area and the majority of those that remain are just empty towers.

Anglesey

A considerable number of tower mills with wheel-and-chain luffing gear and common sails were found here.

These remarks only apply in general, and it is impossible to be dogmatic on any point. Each windmill was constructed individually, and no two are alike in more than the main details. A millwright might incorporate his own ideas, or even travel to a distant county, returning with some idea quite out of keeping

73. SHIPLEY, SUSSEX Restored to working order in 1958, this mill looks much as it did during its working life.

74. NUTLEY, SUSSEX Restoration in progress, 1970.

with the local traditions. There are several examples of this, one being the Lincolnshire-pattern East Blatchington pumping mill, Sussex. Another was Horringer post mill, Suffolk, which had a tailpole fantail.

10 · Windmill construction

Apart from certain design features old windmills have one feature in common—that of the use of large-section timbers, usually of oak. A study of most old constructions, horse carriages for example, reveals that the timbers became of smaller and smaller section as time progressed. These were both strong and light when new, but as soon as a little weakness develops much rebuilding is required. Windmills seem to have undergone this change also, and a study of post mill corner posts will illustrate this well.

Pitch pine is both strong and light, but rots much more quickly than oak.

Given the same degree of neglect, a pitch pine mill will become weak years before its oak counterpart.

Smock mills were usually constructed piece by piece, with the cant posts being erected and the diagonal braces and floor beams fitted into their respective mortises, from the bottom upwards. A suggestion has been made that post mills were constructed piece by piece, with one side girt being put up at a time and the corner post fitted into its mortises on the upper and lower side girts while these were suspended in mid air, no scaffolding being used. A millwright of the old school maintains that such a practice would be impossible, and that the only method of construction was that of raising complete sections of framing. The author used this method when constructing a one-third-size post mill, as the other was quite unrealistic. The crowntree was tied down to the crosstrees at both ends, and first one side and then the other hoisted. An assistant fitted the cross members in the breast and tail as the second side was positioned.

For hoisting, the millwrights probably used a tall pole held upright by guy ropes. A strong block and tackle was suspended from the top; windshafts weighing many tons were raised by such devices. The main timbers were formed either by a hand saw and saw pit or, in older mills, by the use of an adze. The adze was a most useful tool consisting of a sharp blade with a handle about four feet long. A skilled man could use this axe-like tool with extreme accuracy.

It is probable that the millwrights followed the practice of most builders and carpenters and formed the basic timbers in their workshops. These would be brought to the mill site as required. The boards and minor timbers would be cut to size as the work proceeded.

11 · The post-war years

After World War II the decline in the remaining working windmills was rapid. These mills were all worked seriously as part of the miller's business, and as their owners died or a stock became weak they stopped for good. The shortage of timber caused by the war also resulted in several sound mills being destroyed. One method used by millers to keep their mills working was that of obtaining second-hand sails from derelict mills. These were altered as necessary and put to work for a few more years. The supply of second-hand sails gradually dried up, and the only alternative was a new set. The cost usually prohibited this, and after working with only one pair for a while an engine took over.

Engines had been used for auxiliary power since the coming of practical steam

75. A remarkable piece of work. The trestle at NUTLEY which was renewed by a group of enthusiasts.

76. PUNNETS TOWN, SUSSEX Rebuilt from a tower to the condition shown in 23 years.

engines during the eighteen hundreds, and many old photographs show a tall chimney rising from a Victorian brick building beside the windmill. A pulley may often be seen projecting through the side of a post mill or from the tower of a smock or tower mill. This enabled a portable steam engine to be brought to drive the stones in a period of calm. A drive by bevel ring on the great spur wheel was usual in most mills, but some post mills had a pair of iron bevel gears on the stone-spindle.

A few teeth were designed for removal on many brake wheels, which allowed the wallower to be rotated from beneath without revolving the sails. A large number of post mills had the engine drive to a pair or more of stones in the roundhouse.

Steam eventually gave way to gas or oil engines which may occasionally be seen *in situ* to this day. These engines had one very large horizontal cylinder with exposed conrod, big end and crankshaft. The crankshaft carried one or two flywheels of about four feet diameter and a belt pulley. The belt often passed into the mill through a short tunnel or passage which helped to isolate the engine shed from the milling process. The exhaust was discharged through a very large-diameter vertical pipe with a most distinctive chuffing note that could be heard from quite a distance.

The use of these engines has probably prolonged the existence of many mills, and has certainly enabled those that have worked by wind since the war to

77. CROSS-IN-HAND, SUSSEX, as it stood in 1970.

continue. The drives were so varied that they defy description in detail, but the lack of guards around both belts run on points and gears has caused several untimely deaths. Unfortunately, many auxiliary engines have been broken up for scrap, and steam engines are particularly scarce.

The decline in the numbers of working windmills continued, and their scarcity at last provoked an awakening of interest in windmill preservation. The 1950s saw real progress, with several working mills receiving financial assistance when repairs were required. The most usual form of preservation was that undertaken by local and county councils. Beneficial as it is to preserve windmills in any form, these works, although praiseworthy, were frequently of poor quality. In many cases millwrights were not used and, when they were, they were sometimes dictated to by borough surveyors who knew nothing of windmills.

A typical restoration of the mid-to-late 1950s usually concerned itself only with the aesthetic aspect of a local windmill. A set of sails and a coat of paint was all that concerned the restorers, and it was immaterial to them whether a side girt could be half broken through or a cant post in two halves. In the concern for sails insufficient attention was frequently paid to making the mill water-

78. (*Opposite*) HILDENBOROUGH, KENT The mill is shown in the act of falling in 1961.

79. TELHAM HILL, SUSSEX.   80. TELHAM HILL after demolition, 1962.

tight. The rain penetrated to the dry interior timbers which, once exposed, can rot surprisingly quickly.

Councils have been, until recently, the most prolific repairers of mills, but it is rather unfair that they should receive all the blame for what has been done. Some councils have carried out very good repairs which could almost be termed restorations. The secret of these is that they have employed millwrights for all the work and not just for the bare minimum.

About ten years ago the author witnessed the final phase of the restoration of a certain windmill. A millwright had been commissioned to deal with the new sails, which were excellent. The council had repaired the body, using scaffolding which surrounded the mill for some time. The paint had been renewed, but it had been applied to an exterior which had been patched with tin and hardboard. In the tail it was possible to push a finger right through the rotten weatherboarding. Inside the mill the shutters for closing the windows were still lying; they had not been replaced ten years later.

No public body would restore the interior of an ancient castle but leave a leaking roof. Why then should they do what amounts to the same thing to an equally historic monument? The answer is twofold; lack of knowledge and lack of money. If money is short it is well worth restoring the tower or body to a perfect condition, leaving the sails to a later date.

The lesson concerning millwrights seems to be penetrating, but that concerning the use of poor materials is not. Many readers, who have recently observed that the soft pine stock of a local preserved windmill has snapped after

72

only ten years' use, were probably amazed to note that a stock lasted forty years in the past. The answer is simply that the stocks were made of pitch pine or other excellent timber, and stocks of this material should last equally long today. Any council contemplating repairs to a windmill should first consult a millwright as to the correct materials for the job. These should then be specified when tenders are sent out. The use of tenders is not always beneficial, as the cheapest work is not necessarily the best.

Naturally enough most preservers prefer to own the freehold of a windmill, and, with public money in use, local authorities are no exception. The problem is that, with all the sub-committees and official delays involved, the windmill may deteriorate much more than need have been the case. There are at least two cases of windmills collapsing during negotiations which took several years. In more than one instance the council has settled for a very long lease at a very low rent. The standard of repair is becoming better, and it is to be hoped that public bodies will in future concern themselves not only with the exterior, but also with the interior and the small buildings which often surround a windmill. These buildings were often as old as the windmill and were as much a part of the whole as the machinery. Careless repair has led to much machinery being ripped out and many small granaries and barns being demolished.

12 · The question of originality

When restoring a windmill it is of importance that the work is a perfect reproduction of that originally found. There are some obvious examples of very bad reproduction in which the maximum weather is towards the tip of the sail, where the weather is constant or where the sail is just a flat framework. Such careless work is very bad indeed, and is usually the result of work by local builders or council workers who know nothing of the construction of a windmill. This type of work has become scarce in recent years, and it is to be hoped that the offending owners will replace such disgraces as soon as possible. One extreme example of this kind of work carried two clockwise and two anti-clockwise sails!

A more subtle type of mistake is that of changing the type of joint used between the various timbers. This again is uncommon today, but in the past side girts have been recessed into the corner posts, and joined by iron plates, without using a tenon. This is just as strong as the original, but when it is realised that the work may well be examined in five hundred years' time, and believed to be original, it is placed in true perspective. When making a joint or fashioning a piece of timber this fact should be imprinted upon the mind. If the timber to be replaced is missing, then the joints and construction should be as near as can be judged by using local windmills as an example. Should these all have been demolished, then the same type of joint should be used as in other areas. The recording of the design of the various joints has not been carried out very systematically, for many enthusiasts feel that they cannot waste expensive film on such matters. The exposure of such places as the crowntree ends is not common, and when they are so exposed, these probably display a joint so neglected that they do not demonstrate the type well.

Some people argue that if windmills were still in commercial use today the old millwrights would have adopted the materials and methods of the time. This is undoubtedly true, but the fact remains that windmills are a relic of the past, and have been relegated to the ranks of castles and old sailing ships. Like vintage cars, windmills can still be used in the original manner, but this is no reason why they should not be maintained to high standards of condition and originality.

In Holland, where windmills were very common, a special design of sail evolved just before World War II. This consisted of a very long stock which also replaced the whip; the sail bars were mortised through in the normal way and the sails built up after the stock had been erected. This stock was often built out of steel plate. Cranbrook mill, Kent, was restored in 1960, and in spite of a very large sum of money being expended was completely ruined by the addition of this type of sail. Apart from the practical disadvantage that the sails are now very close to the mill tower, this type of degradation should be

81. ASH, KENT Blown down, 1953.

82. OUTWOOD smock mill after being blown down, 1961.

avoided at all costs, for the mill is no longer typical of either England or Holland.

The finest way to preserve a mill is to keep it at work, but as this is not possible in a great number of cases, certain precautions have to be taken which are quite excusable. A windmill like that at Chillenden, which stands alone in a field, often receives the full attention of vandals. The pumping mill at Warbleswick, Suffolk, was completely burnt out a few years ago through the efforts of these people. To avoid damage of this type the lower windows are often boarded or bricked up and the door very securely locked.

The sail shutters are frequently removed to lessen wind-resistance in a tail-wind, a variation being the use of bare shutter frames when these were covered with canvas. Another excellent idea which is becoming more frequent in use is that of having only a few shutters in the sails, the post mill at Saxtead Green, Suffolk (Fig. 66), being an example. The owner of Barham smock mill, Kent (Fig. 72), used an equally sensible arrangement in which he had a full complement of shutters only for about two-thirds of the way out from the heel. The last few bays were empty.

The fan is often erected in skeleton form, also to lessen wind-resistance, an example being at West Kingsdown, Kent. This last mill illustrates the increasing belief in restoration to an original form, for the fan stage was completely missing before work commenced. The excellent mill at Argos Hill, Sussex, has recently had the tailpole fantail replaced after it had lain neglected in the field for many years. The fan here also is of skeleton pattern, for with the mill unattended it is naturally not required for turning purposes.

83. CROWBOROUGH, SUSSEX All that now remains of the post mill which stood here.

84. THORNHAM MAGNA, SUFFOLK Last worked 1942. Demolished about 1962.

With such an example as this it is to be hoped that other public bodies will follow by restoring their mills, not only externally but internally also. Excellent examples of restoration have been achieved by private individuals forming committees or trusts which raise money by means of public subscription, fêtes or other means. The result is often very original, and may even be used for work.

ENTHUSIASTS

To most windmill enthusiasts a shell restoration is of little interest, for they are concerned with originality and the recording of windmills as they were constructed by millwrights and used by millers. A derelict Cambridgeshire smock mill is far more likely to receive a visit from such people than a hollow tower with a recent cap like Halnaker in Sussex.

Most dream of owning a real windmill and even working it, but usually have to be content with a model, which either spins merrily on the lawn, or is used to demonstrate windmill construction in the sitting-room. One such enthusiast takes this to an extreme and even has salt-cellars made like small windmills. Some attain their ambition and either own a preserved example, or restore a derelict mill to working order. Others with rather less finance join together in groups to carry out the complete restoration of a local mill.

Some quite remarkable work has been achieved in this way, for these people not only know what the result should be like, but also do not charge for their many hours of work. A complete rebuild can thus be found to have cost only a few thousand pounds.

RESTORATION TO WORKING ORDER

An outstanding piece of work was commenced by Mr. Dallaway on Punnets Town mill, Sussex (Fig. 76), in 1947. This smock mill had been left as an empty tower by his father, the cap had been removed, and the interior was used for storing cattle-feed in connection with the family milling business. The tower was struck by lightning, and the repair of this damage started what was to become a complete restoration. The cap frame was rebuilt as a circle, the trees being hauled out of the woods and sawn up by Mr. Dallaway himself. The cap is of a domed type with a covering of sheet aluminium. To facilitate repair and repainting, a small gallery was built around the cap. A Sussex-type fan stage was constructed at the rear.

The windshaft came from Staplecross smock mill, when this was pulled down. This was hoisted into position and a fine brake wheel built. A short upright shaft drives to a clasp-arm great spur wheel which meshes with wooden stone-nuts. There is a third wooden pinion which takes the drive through a pair of wooden bevel gears to an oatcrusher.

The fan drives through an assortment of bevel gears to a specially designed skew gear which meshes with the rack, which fortunately still remained. A pair of common sails have been erected and another pair is to follow shortly. The difficulty of manufacturing the wooden gears and erecting the heavy parts is made the greater by its having been a single-handed effort. As this restoration may be considered the forerunner of all modern complete rebuilds, the use of proprietary gears in the fan mechanism is quite acceptable. Most of the major restorations of this type have started with a complete mill which has very rotten timbers; this rebuild is made the more exceptional by having originated from an empty tower.

During 1958 two interesting restorations were started which, while undertaken by the Ministry of Works and a group including a county council, were of a type hitherto unknown; the complete restoration of windmills to working order.

Saxtead Green post mill Suffolk (Fig. 66) was completely dismantled, leaving only the trestle and roundhouse standing. Every timber which was not in perfect condition was replaced by the same type of timber as was used originally, and a new crown tree was fashioned from that of Wetheringsett mill which had recently been dismantled. The standard of workmanship was very high, and the mill is now in new condition. Assuming that the maintenance is of the same standard, there should be no need for major repairs to the mill body for many hundreds

86. WESTON, SHROPSHIRE Derelict, but contains a compass arm brakewheel and wallower.

85. HAWKINGE, KENT Blown down 1963.

87. NORTHWOOD, KENT Demolished 1960. Now a housing estate.

88. BEXHILL, SUSSEX Downs Mill, collapsed mid 1960s.

89. A rare sight today, one mill seen from another. WOOLPIT from DRINKSTONE post mill.

90. SHIREMARK, SURREY Now very derelict.

of years. A restoration of this magnitude was made possible by the mill being owned by the Ministry of Works, who have very deep pockets, and the fact that Mr. Jesse Wightman the millwright lived locally and was able to supervise the entire work.

The windmill is in first-class working order but, sad to relate, it is not allowed to grind, and the sails are only fitted with enough shutters to allow them to idle when there is a good breeze.

Shipley smock mill, Sussex (Fig. 73) was restored about the same time as a memorial to the writer Hilaire Belloc. The work was undertaken by the millwrights E. Hole & Son, and was exceptional in that a decision was taken that the mill should be made actually to grind corn. Built in 1879, this is a very modern mill with much iron machinery, and was in fact in quite a reasonable state of repair. The stocks were replaced, and repairs carried out to the cant posts, sails, cap frame and boarding, as well as to minor items like the stage. The generally good condition of the mill is indicated by the fact that the bill came to only a little over £4,000. When it is considered that the mill was made to grind and that it had last been used in 1926, this is not expensive.

Mr. Powell, the son of the last miller, used the mill a good deal after the restoration, and in fact ground sixty tons of grain during one winter. Since his untimely death a few years ago a little grinding has been done, but only for demonstration purposes. The sails may be seen idling two days a month during the summer, when the interior can be inspected for a small charge.

The last few years have seen a great increase in restorations, which not only restore the outside but renovate the interior also. These restorations are usually undertaken by groups of private individuals who are also willing to assume the positions of millwright, carpenter and labourer. The work is all free and so the only charges are for materials and such items as the hiring of cranes or special tools. An appeal for funds frequently produces some money quite quickly, which allows work to start before the weather causes any further damage.

Wrawby post mill, Lincolnshire, was restored by a group of enthusiasts who rebuilt the mill completely after dismantling virtually the entire structure. Post mills are extremely rare in the North East which makes this restoration of great importance, for not only is the mill restored, but it is also in frequent use.

Another mill in constant use is that at Over in Cambridgeshire, which has been restored by its owner in his spare time. This was a single-handed piece of work which deserves great credit.

The windmill at Nutley in Sussex (Fig. 74) is a small old post mill of considerable age which last worked during the 1920s. It stood with the battered remains of its spring sweeps until the owner of the estate, upon which it stands, had a set of bare sail frames erected in the early 1930s. The body was also

91. SIX MILE BOTTOM, CAMBRIDGESHIRE
An interesting East Cambridgeshire post mill.

92. A nearby clockwise smock mill at FUL-
BOURN, CAMBRIDGESHIRE.

underpinned in the breast as there was a damaged side girt. More work was carried out in the mid 1950s, which included the strengthening of some main timbers.

The end of 1969 saw the trestle having decayed to such an extent that the post had dropped down from the crowntree and the entire weight was being taken by the underpinning. The two corner posts in the breast had almost rotted through, and the breast beam which they supported was carrying the full weight of the sails.

An agreement was reached between a private group and the owner, and the tricky job of removing the sails without bringing about the destruction of the entire mill was accomplished. An appeal was launched and work proceeded with great speed. The owner did not wish the mill to be dismantled and rebuilt, which meant that the whole body had to be supported while many main timbers were replaced. A remarkable achievement was the remaking and re-placing of the crosstrees and quarterbars. This difficult work was carried out at weekends during the winter months and was completed in a very short time.

A pleasing aspect of this restoration is that any timbers which are usable are being retained and if the joint has become rotten a new end is spliced on. The final result should be as near to the original state as possible, and yet the whole building will be as good as new. When entering a mill like Saxtead Green there is an inevitable feeling that it is a new mill. This should not be the case at Nutley, the reconstruction of which is an interesting attempt to combine the best of both worlds.

13 · Working a windmill today

Today there is only a small handful of windmills which are still used in conjunction with a milling business. Those at North Leverton, Nottinghamshire (Fig. 102), Pakenham in Suffolk (Fig. 21) and Stelling Minnis in Kent (Fig. 12) are still the mainstay of their owners' businesses. That at Drinkstone (Fig. 100) is an adjunct to the owner's business, and is really run for pleasure rather than commerce. The same applies to Alford, in Lincolnshire (Fig. 101) where the tower mill is owned by a firm from a nearby village who run it occasionally. The majority of the remainder are used only for demonstration to the public, two of the exceptions being Punnets Town (Fig. 76) and Over.

Windmills are no longer a commercial proposition, and behind every working mill there is an enthusiastic owner. Without these people there would be no windmills working in this country today. Farmers are able to buy their animal-feeds from big milling companies who deliver ten tons at a time from as far away as forty or more miles. Such companies are often able to supply the feed cheaper than the local miller, and with farming itself under severe economic pressure, inevitably trade has drifted away. Even cattle- or pig-food has special additives nowadays, which can often only be contained economically in feeds produced in vast bulk. Many small millers are agents for these large companies, but none the less are frequently closing their doors for good.

A problem which could well face windmillers in the future is that of disposal of the finished product, although there may still be scope where specialist feeds

93. WILLASTON, CHESHIRE There is wheel-and-chain luffing gear and a cross instead of a poll end.

94. BURWELL, CAMBRIDGESHIRE, shortly after ceasing work for good. The cap and fantail are typical of the area.

are concerned. Very large firms are usually very reluctant to produce any special or small batches, and this could possibly react in the small miller's favour.

Stone-dressing is a problem also, for those who are really skilled are becoming few indeed. Since the death of Mr. Powell, the stones at Shipley have not been dressed, and it is to be wondered who will do this when it is needed. This problem must face most groups of enthusiasts when they decide to grind with their newly restored mill. The problem of replacing worn-out stones is not great, for there are hundreds of stones throughout the country which remain from demolished mills of all types. Stones do not wear very quickly anyway, and should easily last for ten years even in constant use.

To the problems of wear and of dressing there is one simple remedy, namely composition stones. These are often one-piece and are composed of carborundum or other very hard chippings mixed with cement. The surface is very rough and the fine stitching is not required, leaving only the furrows to be deepened and the surface levelled when the stones are lifted. Since they are composed of extremely hard chippings, dressing is only very infrequently required.

A small milling firm which owned two water mills and a one-time steam mill used a most interesting system. An ordinary millstone was turned face up and fitted with an extra iron band around the circumference. This was left a few inches proud, and a cylindrical piece of tin was fitted in the eye, this also being left a few inches proud. A mixture of cement powder and hard chippings was produced to which a quantity of spirits of salt was added. After thorough stirring the resultant cement was spread carefully over the surface of the stone. The stone was left in this condition until the cement was absolutely dry, when the furrows were marked out and the surface levelled where necessary. The iron band was then punched down below the level of the stone face.

To cut the furrows and to carry out all dressing operations a tool known as the Kango hammer was used. This is best described as an industrial drill which, while driven by electricity, operates on the principle of a pneumatic road-drill. A number of different bits are provided, and in skilled hands the drill can be used to great advantage. A Kango hammer would certainly appear to be the answer to many a miller's dressing problems.

Minor repairs are usually carried out by the miller himself. These normally include repainting, which used to be undertaken every three years or so, although nowadays there is frequently a longer gap. When painting a white post mill, like Keymer, Sussex, about three hundredweight of best lead paint was required. The mill body was painted from a cradle which was suspended by ropes passed over the mill roof and down to the ground on the far side. The cradle was entered either by a ladder or from a window.

95. BILLINGFORD, NORFOLK Shown soon after it stopped for good in the early 1950s.

96. SYLEHAM, SUFFOLK A few years after it last worked by wind.

Tar was frequently used to preserve the boarding and was very useful although very messy to apply. This does have the disadvantage of having to be heated before application, but is excellent in its sealing properties and probably well repays its use. Many Kentish smock mills were tarred.

Those who have worked a windmill will agree that it is a fascinating experience, and perhaps even those who have not had the pleasure of having the mill under their own control, but have visited working mills, will know something of this feeling. It is not surprising therefore that, despite the problems of maintenance and the cost of upkeep, people still continue to work windmills. People sail boats on the sea for pleasure, and there is certainly a great affinity between the two machines.

Naturally the greatest similarity is between an old post mill and a small sailing-boat, for the post mill becomes alive when in motion, and if the joints are rather worn the whole mill shakes and rolls when running fast in a stiff breeze. A tower mill is a far more reserved and stately affair, for when inside there is no apparent sensation of movement in the tower, and only the rumble of the runner stone, the faint clacking of the shoe and the rattle of the sack-hoist give any clue that the mill is at work.

As you enter a Lincolnshire-type tower mill the immediate sensation is one of warm dustiness, for there is a thin film of flour dust on most surfaces. Sacks are stacked against the walls, some full of grain waiting to be ground and others containing meal ready for despatch to the various local farms. To one side a broad ladder gives access to the floor above. The treads are deeply worn with the passage of hundreds of pairs of feet. A rope helps the visitor to steady himself as he climbs to the next floor.

This is the spout floor, and three bins stand together in the centre of the floor with wooden spouts descending to them from the floor above. Dust is in the air, for the miller is grinding barley which is always rather dusty. To try to reduce this he has left one of the two loading doors open, and this also helps to make the floor much lighter than the one below. There are always two loading doors in order to facilitate the loading of wagons or lorries, whichever way the sails are facing. The bins are highly polished by the friction of centuries of sacks, and into one a steady flow of barley meal is pouring. When handled this has a sweet smell and a yellow tinge. The miller is feeling the meal between his thumb and fingers and is adjusting a large wing-nut on the tentering gear above his head. In the ceiling the lower surfaces of the bed stones are visible, as are the governors driven from the stone-spindle.

The next floor up is the stone floor with one pair of stones rumbling beneath a gently rising dust cloud. The shoe is tapping away beneath the hopper, in which the grain is observed to be constantly on the move as it flows out of the almost submerged end of the spout from the bins on the floor above. The great spur wheel is revolving overhead, and one stone-nut is in gear while the others are leaning back on their quants in the disengaged position. The wooden teeth are well greased, and there is hardly any sound as they mesh with the iron stone-nut. Through one window the backs of the white-painted sails can be seen flashing past.

The bin floor is quite small, for the tower has tapered considerably and almost the entire area is taken up by the bins. There is just room to stand beside the sack-trap and notice the shining sack-chain disappearing down through the much-worn holes to the floors below. The iron upright shaft is revolving slowly in the centre.

Another ladder leads to the dust floor which is even smaller than the last, and contains little besides the upright shaft and wallower with the sack-hoist ready for engagement with its lower side. Overhead the cap rises, with the brake wheel and windshaft turning slowly. Again there is little noise from the meshing of the wooden brake-wheel teeth with those of the cast-iron wallower.

The wind is whistling around the cap, and there is quite a draught caused by the wind coming up under the petticoat (the name given to the ring of boards around the base of the cap). The wind can be heard in the sails, and when a heavy gust strikes them the whole cap vibrates. A narrow portable ladder gives access into the cap itself, and when this is climbed the motion of the mill is at once apparent. The door to the fan stage is open and allows the only light to penetrate, for there are no windows on the dust floor, and the fan gears can be seen engaging with the thickly greased curb. A tub of grease for lubricating these and the neck and tail bearings is resting on the cap timbers. Beside them there is a large oil-can for oiling the fan bearings.

Stepping over the curb, it is possible to go out on to the small fan stage. This can be felt moving with the motion of the sails, and becomes even less steady when the fan suddenly starts to turn the cap. A ladder climbs towards one of the cross-braces of the fly posts from which the hand-rail of either fly post may be reached. To the visitor the prospect of climbing up the wooden blocks of the fly posts is distinctively uninviting, for the wind can now be felt tugging at him as he climbs the short ladder and emerges into the air-flow. Turning to face the wind, the patent sails are seen turning on the far side of the beautiful ogee-shaped cap. As each reaches its zenith the air-flow is momentarily cut off from the observer, to resume as the sail rattles its way earthwards.

From a spot such as this sixty years ago, it might be possible to see as many as forty other windmills all turning busily in the wind. Today, you may be fortunate to glimpse one empty tower on the skyline. Not only have the windmills gone, but so have the men who worked them, and these days it is as much a privilege to meet an old miller who started work in Edwardian times as it is to stand on the fan stage of a truly working windmill.

THE FUTURE

Regrettably, the days of the windmill seriously grinding are passing, and it may only be a matter of time before all the mills which remain at work do so only for demonstration. This is sad, for the genuinely active mill has a special atmosphere which is not present in a mill whose sails are only allowed to idle while dozens of casual visitors, screaming children and all, wander over the dust-free floors. How much better it is to see windmills restored fully and their sails turning than to see the shell restorations of the late 1950s!

There is a definite future for the windmill in this new, less active role, and there are at least three which have been restored from dereliction to working order for the enjoyment of visitors. These efforts are most praiseworthy, but it is

97. The derelict post mill at WOOLPIT, SUFFOLK Last worked in 1953, it collapsed in the mid 1960s.

98. DRINKSTONE post mill, SUFFOLK The last post mill to work in the county, this is also the last corn mill to work regularly with common sails.

a pity that mills in good order, which would require little more than maintenance, have disappeared from the active list during the same period.

Recent years have seen a notable increase in leisure hours and incomes. This
has resulted in people taking up interests and hobbies at an increasing rate.
Some collect stamps or have other indoor interests, while the more energetic
restore vintage cars or old buildings. It is presumably from this last group that
the increasing number of windmill enthusiasts is coming. Assuming that the
pattern does not change, there seems a strong possibility that this is only the
thin end of the wedge, and that the 1970s will produce more and more restorations to working order. Many of these will be in the hands of private groups
or individuals, and the day may yet come when the owner returns home from
a five-day week at his office to run his windmill over the weekend.

Windmills to be seen

A list of interesting windmills, most of which have been restored and can be viewed from the road.

The purpose of this list is that of providing the reader, who has no previous knowledge of windmills, with the names of an interesting selection which are externally reasonably complete. Most of these have been restored and may be seen from the road. The great majority are privately owned and for this reason are not accessible to the public. This same ruling also applies to those in public ownership, for although most councils are quite willing to show people around, provided an appointment has been made, it is hardly fair for the casual visitor to bother them in this manner.

The original intention was that this section should cover all those counties which contain windmills, in approximately equal proportions. A preliminary survey of the situation, however, at once made it quite obvious that some counties contain far more restored windmills than others. A county like Kent, for instance, is bristling with restored mills, many of these being of excellent quality, and it would have been senseless to have excluded these just to keep the proportion equal to that of Lincolnshire.

Those mills marked * are believed, at the time of writing, to be open to the public.

ALFORD Lincolnshire. Five-sailed tower of typical Lincolnshire design. Worked occasionally.

ARGOS HILL Sussex. Interesting post mill owned by local council. Unusual half-extension at the rear. Recent restoration included the re-fitting of the tailpole fantail, making this one of only two mills remaining with this fitting. The other mill stands at Cross-in-Hand, just across the valley.

ASHTON Somerset. Tower mill rather typical of the area, recently restored.

AYTHORPE RODING Essex. Restored post mill with ladder fantail.

BARDWELL, Suffolk. Tower mill without sails or fantail, but interesting for beehive cap and condition.

*BERNEY ARMS Norfolk. Very tall and magnificent drainage mill, restored.

BIDSTON Cheshire. Restored tower mill, displaying typical local cap.

BILLINGFORD, Norfolk. The last corn mill to work in the county, now restored.

BOCKING Essex. Restored post mill.

BOURN Cambridgeshire. The oldest windmill in the country. Small open-trestle post mill which was restored some years before World War II.

BRILL Buckinghamshire. Ancient post mill, preserved, but the ladder is missing.

BRIXTON London. Tower mill, wheel and chain luffing mechanism, and recently restored.

BURGH-LE-MARSH Lincolnshire. Clockwise, five-sailed tower mill, which ceased working a few years ago.

BURWELL Cambridgeshire. Tower mill which ceased to work during the late 1950s. Typical cap and fantail.

CAPENHURST Cheshire. Disused, but fairly complete tower mill, with the remains of three common sails.

CHILLENDEN Kent. Interesting, preserved post mill with open trestle. Regrettably much machinery removed and adjacent granary destroyed.

CLAYTON Sussex. Probably the most famous windmills in England. One tower mill without machinery or sails, but complete with typical cap and fanstage. One post mill with sails, but original tailpole fantail missing. Moved to present position from Dyke Road, Brighton. Very strongly built, some machinery removed.

CROSS-IN-HAND Sussex. Last windmill to work seriously in the county. Recently damaged by falling stock. Side girts and frame generally in poor condition, therefore unlikely to work again without great expenditure.

DRINKSTONE, Suffolk. One working clockwise post mill, one very interesting, but empty smock mill. Post mill of great age, extended in both the breast and the tail. Uses the only common sails now at work on a corn mill.

FRAMSDEN Suffolk. Post mill with ladder fantail. Raised to present height at great cost. Restored.

FRISTON Suffolk. Pronounced the finest remaining post mill in the county by a local millwright, this mill sadly ceased to work during the early 1960s. Superb mill on three-storey round-house.

GAYTON Cheshire. Very old tower mill, regrettably derelict, but possesses wooden poll end.

GREAT CHISHILL Cambridgeshire. Small open-trestle post mill with fantail attached to the ladder. Very complete, good order.

HADDENHAM Cambridgeshire. Large tower mill with typical cap and fantail.

HECKINGTON Lincolnshire. The only eight-sailed mill remaining. Well preserved by local council.

HERRINGFLEET Suffolk. Drainage smock mill. Four common sails. Preserved.

HIGH SALVINGTON (near Worthing), Sussex. Extensively rebuilt post mill. Four common sails erected.

HOLTON Suffolk. Post mill with two common and two shuttered sails. Ladder fantail.

HORSEY Norfolk. Well-restored drainage mill with typical cap and fantail.

ICKLESHAM Sussex. Post mill with roof fantail. Four sails recently fitted.

KEYMER Sussex. Preserved. Very large floor area, making this, surprisingly, one of the largest post mill bodies in the south. Very early wrought iron universal joint in sack hoist drive.

MADINGLEY Cambridgeshire. Small post mill, with two common and two shuttered sails. Staggered teeth on brake wheel and stones on hurst frame. Moved to present site between the Wars.

*MARGATE Kent. Smock mill, recently restored.

MEOPHAM Kent. Smock mill restored some years ago to present condition.

NORTH LEVERTON Nottinghamshire. Most interesting, working tower mill. Repairs carried out a few years ago enabled this mill to continue to work. Very good condition.

NUTLEY Sussex. Small post mill which last worked with spring sails. Very thorough reconditioning in progress.

99. (*Opposite*) A typical Lincolnshire tower mill, ALFORD, LINCOLNSHIRE is still used occasionally. The last multi-sailed mill in use.

*OUTWOOD Surrey. Post mill erected in 1665. Four spring sails. Worked for demonstration purposes.

OVER Cambridgeshire. Tower mill restored and worked regularly by present owner.

PAKENHAM Suffolk. Working tower mill. Good order and in constant use.

PITSTONE GREEN Buckinghamshire. Post mill, restored, contains date 1627.

*POLEGATE Sussex. Tower mill. Considerable restoration carried out, including what amounted to the complete renewal of the cap.

PORTLAND BILL Dorset. Two most interesting, if almost bare, tower mills. One contains a wooden windshaft with wooden poll end.

PUNNETS TOWN Sussex. Smock mill, completely restored by its owner.

REIGATE Surrey. (1) Post mill on Reigate Heath, recently renovated. (2) Tower mill on Wray Common, sails and fantail, but regrettably machinery all removed.

ROLVENDEN Kent. Restored, four common sails fitted, but ladder not replaced.

SANDWICH Kent. Much restoration carried out by Mr. V. G. Pargeter, including the fitting of two sails from Wingham tower mill. Now preserved by public body.

*SAXTEAD GREEN Suffolk. Typical Suffolk post mill, completely rebuilt by Mr. J. Wightman on behalf of the Ministry of Works.

*SHIPLEY Sussex. Late smock mill renovated to working order by E. Hole and Sons on behalf of a preservation society.

SIX MILE BOTTOM Cambridgeshire. Interesting post mill standing just off the main London to Newmarket road.

STANTON Suffolk. Disused post mill.

STELLING MINNIS Kent. The last working windmill in Kent. This mill is both small and of quite recent construction. Presently working with two sails only.

STODMARSH Kent. Possibly the only remaining example of the hollow-post drainage mill.

STONE CROSS Sussex. Tower mill of relatively recent construction. Some repairs carried out.

STRACEY ARMS Norfolk. Restored drainage mill.

SWAFFHAM PRIOR Cambridgeshire. Preserved tower mill. Remains of smock mill nearby.

SYLEHAM Suffolk. Post mill with two sails. One of the last post mills to work in this county. Now disused.

TADWORTH Surrey. Large post mill not far from A23 main road. No sails or machinery. Damage caused by bomb during the last War, repaired when hostilities ceased.

THORPENESS Suffolk. Post mill moved to present site and used to pump water. Now disused. Peculiar modern, square round-house.

THURNE Norfolk. Famous, preserved, pumping mill.

WEST BLATCHINGTON Sussex. Preserved hexagonal smock mill. Very rare mounting for sails, consisting of a cross and short stocks.

WEST KINGSDOWN Kent. Restored smock mill. This was once one of a pair, but the post mill which stood close by was burnt down. Both mills had two common and two shuttered sails.

100. (*Opposite*) NORTH LEVERTON, NOTTINGHAMSHIRE This mill is in constant use and may be seen at work on any windy day.

WITTERSHAM Kent. Preserved post mill. All machinery, apart from the brake and tail wheels, removed. Windows modern, strange gallery attached to one side.

WICKEN FEN Cambridgeshire. Small pumping mill used by the National Trust.

WILLASTON Cheshire. Tall tower mill, without sails, but with cap.

WILLESBOROUGH Kent. Large smock mill overlooking the bypass. Repaired.

WILLINGHAM Cambridgeshire. Smock mill with ogee-shaped cap. Two sails remain. Last smock mill to work in the county. Last worked during the early 1960s.

WIMBLEDON COMMON Surrey.

A unique design of windmill which, although often described as composite, seems really to be more closely related to a smock mill with an over-large cap.

WOODCHURCH Kent. A preserved smock mill, fitted with a new set of sails about ten years ago. The remaining mill of a pair.

*WRAWBY Lincolnshire. Post mills are of great rarity in the North Eastern counties, a fact which makes the restoration of this mill of even greater merit than might have been the case. The rebuilding now completed, this mill is worked regularly for demonstration.

101. CLAYTON, SUSSEX External and internal views of the cap.

Glossary

Terms included in the text and other synonymous or useful expressions. Figures in italics are page references.

ANNULAR SAIL A ring-like sail with radial shutters similar to a modern iron windpump, *36*.

APPOLD TURBINE A type of pump driven by marsh mills.

BACKSTAYS Brace the frame from the stock, *29*.

BAY The space between two sail bars, *31*.

BED STONE The lower stationary mill stone, *44, 45*.

BELL ALARM Warns the miller when the grain is low, in the hopper, *42*.

BIN Contains the grain in the upper floors of the mill, *41, 42*.

BIN FLOOR Contains the grain bins.

BLUE STONE Imported from Germany; one-piece mill stone, *44*.

BODY Upper wooden part of a post mill, *12*.

BOLTER Early type of flour-dresser, *49*.

BOLTING CLOTH Cloth used in a bolter.

BRAKE Operates on the rim of the brake wheel to stop the sails, *16*.

BRAKE LEVER Actuates the brake.

BRAKE ROPE Enables the brake to be operated from any floor.

BRAKE WHEEL Largest gear wheel in the mill, situated on the windshaft, *16*.

BREAST The front of the mill, *16*.

BREAST BEAM Supports the neck bearing of the windshaft, *16*.

BRIDGE TREE Supports the lower end of the stone spindle, *46*.

BRIDGING BOX Adjustable stone spindle bearing on the bridge tree.

BUCK Suffolk post mill body.

CANT POST Corner post of a smock mill, *22*.

CAP The movable top of a smock or tower mill, *19*.

CAP CIRCLE Circular sub-frame from which the cap rafters rise on some mills, *28*.

CAP SHEERS Main lengthways timbers of a cap, *19*.

CLAMPS Pass either side of a poll end to strengthen some stocks.

CLASP ARM WHEEL The arms form a square which grips the shaft, *53*.

CLOTH SAILS The same as Common Sails.

CLOTHS These are spread on common sails.

COCK HEAD The rounded tip of the stone spindle, *46*.

COLLAR Steadies the body of a post mill on the post, *13*.

COMMON SAILS Early cloth-covered sails (cloth sails)

COMPASS ARM WHEEL The arms are mortised through the shaft, *50, 53*.

COMPOSITE MILL The body of a post mill is mounted on a short tower in the same way as a tower mill cap, *28*.

COMPOSITION STONES Manufactured from very hard chippings and cement, *82*.

CROSS Alternative method to poll end for attaching the sails, *34*.

CROSSTREES Cross over one another to form the bottom of a post mill trestle, *12*.

CROWNTREE The main beam of a post mill body; carries the bearing upon which the whole mill turns, *12*.

CULLEN STONES The same as Blue stones, *44*.

CURB The ring on the top of a smock or tower mill body upon which the cap turns, *19*.

DAMSEL Vibrates the shoe in an underdrift mill, *43*.

DEAD CURB The cap skids round on iron or brass pads without the use of rollers, *19*.

DOUBLE SHUTTERED SAILS Shutters on both sides of the whip.

DRESSER General term for bolter or wire machine, *49*.
DRESSING The furrows cut on the face of the stone, *44*.
DUST FLOOR Floor just below the cap.

EYE The hole in the centre of the runner stone, *44*.
EYE STAFF Short staff used for checking the level of the area around the eye, *45*.

FAN A small set of sails positioned at right angles to the main sails, *37*.
FAN BRACES Brace the fly posts from the cap.
FAN SPARS The same as fly posts.
FAN STAGE A stage at the rear of the cap to provide access to the fantail.
FAN STAR Iron hub of the fan, *37*.
FANTAIL Turns the sails to face the wind automatically, *19, 37*.
FLY POSTS Upright posts which support the fan, *37*.
FRENCH BURR A built-up mill stone used for flour, *44*.
FURROWS The main grooves in the grinding face of a mill stone, *44*.

GALLERY Platform around the cap or tower of a smock or tower mill, *37*.
GATE A slide found in some shoes to restrict the flow of grain.
GOVERNOR Maintains the correct gap between the mill stones, *47*.
GRAIN CLEANER Machine for cleaning grain, *50*.
GRAFT SHAFT A wooden shaft to which an iron extension has been added.
GREAT SPUR WHEEL The large spur wheel which drives the stone-nuts, *23*.
GRIST Term for meal used for animal food.
GUDGEON Iron pin projecting from a shaft to form a bearing.

HACKLE PLATE Prevents dirt entering the bearing in the bed stone.
HEAD-SICK The mill leans forwards.

HEAD WHEEL Brake wheel.
HEMLATH Joins the tips of the sail bars, *29*.
HOLLOW POST MILL The drive passes down through the post, and drives the machinery in the roundhouse, *28*.
HOPPER Contains the grain just above the stones, on the stone vat, *42*.
HORSE Frame that supports the hopper and shoe, *42*.
HURST Frame in the breast of of some Midlands post mills which carries the stones, *48*.

JACK STAFF Used to check the vertical position of the stone-spindle, *45*.
JOG SCRY Inclined trough with sieves in the bottom to grade flour, *50*.
JUMPER Jog Scry.

LANDS The raised parts between the furrows of a mill stone, *44*.
LANTERN PINION Early type of bevel gear, *50, 53*.
LATHS Lengthways bars of a common sail.
LEADING BOARD On the leading side of a single-shuttered sail, *29*.
LIVE CURB The cap turns on rollers, *19*.

MACE Driven by the quant or stone-spindle, *45*.
MEAL BIN Receives the ground meal from the stones.
MIDDLING Kentish term for a stock.
MILL BILL Tool for dressing the stones, *44*.
MULTI-SAILED A windmill with more than four sails.

NECK The bearing behind the poll end, *16*.
NECK BEARING The neck of the windshaft turns on this, *16*.

OVERDRIFT Stones driven from above.

PATENT SAIL The shutters are

reefed from within the mill, *32*.

PEAK STONE A one-piece millstone mined in England, *44*.

PETTICOAT Vertical boarding around the lower part of a cap or post mill.

PICK Pointed mill bill used to dress the stones, *44*.

PIERS Take the weight of the crosstree ends, *12*.

PINTLE The bearing at the top of the post, *12*.

PIT WHEEL Drives the scoop wheel of a drainage mill, *56*.

POLL END Used to attach the sails on many mills (canister), *16*.

POST The body of a post mill turns on this, *12*.

POST MILL The entire body turns on a main post.

PRICK-POST Vertical beam in the breast of a postmill.

PROOF STAFF The staff is checked against this.

QUANT Drives the stones in an overdriven mill, *42*.

QUARTERBARS Diagonal supports of the main post, *12*.

RACK Gearing around the curb of a smock or tower mill, *19*.

RED OXIDE Used on the staff to test the level of the stone face.

RED STONE Type of millstone used in the north-west, *44*.

ROLLER-REEFING SAILS These use roller blinds instead of the shutters of a patent sail, *34*.

ROUNDHOUSE Building to protect and provide storage around the trestle of a post mill, *18*.

RUBBING BURR Piece of hard stone used to rub off the high places of a millstone, *45*.

RUNNER STONE The upper revolving millstone, *44*.

SACK CHAIN Raises sacks to the upper floors, *48*.

SACK-HOIST Used with sack chain, *48*.

SACK SLIDE On some post mill ladders, used to lower sacks.

SAILS Utilise the wind-pressure to drive the machinery, *29*, *36*.

SAIL BACK Strong sail whip used with a cross, *34*.

SAIL BARS Crosswise bars of the sail frame, *29*.

SAMSON HEAD Iron bearing fitted at bearing of post and crowntree, *12*.

SCOOP WHEEL Raises water on marshes to a higher level, *56*.

SHEER BEAMS Pass fore and aft under the spout floor of a post mill floor, *13*.

SHOE Feeds grain from the hopper into the eye of the stone, *42*.

SHUTTERS Open and close in the same manner as a Venetian blind in the bays of spring and patent sails, *31*.

SHUTTER BAR Connects the shutters, *31*.

SIDE GIRT On either side of a post mill these each take half the weight, *13*.

SINGLE-SHUTTERED SAILS Have shutters on the trailing side only.

SKIRT The outer section of a millstone.

SMOCK MILL A wooden tower with a cap, which turns to face the wind, *19*.

SMUTTER A type of vertical grain cleaner which removes a fungus from grain, *50*.

SPIDER Operates the shutter bars of a patent-sailed mill, *32*.

SPRATTLE BEAM Carries the upper bearing of the upright shaft. Also the upper bearings of the quants, *19*, *46*.

SPRING SAILS A shuttered type of sail controlled by a spring, *31*.

SPRING-PATENT SAILS A rare type of patent sail which has a spring in the striking gear of each sail to provide individual regulation.

STAFF Used to test the surface of the stone for high spots, *44*.

STAGE Gallery around a smock or tower mill.

STAVES The equivalent of teeth in a lantern pinion.

STEELYARD A long level for tentering the stones, connected to the governor, *46.*

STOCK The main timber which supports the sails when a poll end is used, *16, 29.*

STONES Grind the grain, *41.*

STONE CASING Stone vat.

STONE-DRESSING The act of recutting the furrows in the grinding face of the millstone, *44.*

STONE FLOOR The floor upon which the stones are situated, *18.*

STONE-NUT The gear that drives the stone, *16, 41.*

STONE-SPINDLE Supports the runner stone, *45.*

STORM HATCH Allows access to the poll end from within the mill.

STRIKING CHAIN Operates the striking gear.

STRIKING GEAR Operates the shutters of patent sails, *32.*

STRIKING ROD Passes from end to end of the windshaft to actuate the shutters of patent sails, *32.*

SUBSTRUCTURE Term for the trestle of a post mill.

SUNK POST MILL The trestle is buried in the ground, *28.*

SWEEP Southern name for a sail.

SWEEP-GOVERNOR Device for regulating the speed of the sweeps.

TAIL The rear of a windmill.

TAIL BEAM Supports the tail bearing of the windshaft, *16.*

TAIL BEARING The bearing at the tail of the windshaft, *16.*

TAILPOLE Turns many early mills to face the wind, *12.*

TAIL WHEEL Smaller than the brake wheel, this drives the stones in the rear of a post mill, *18.*

TAIL WIND A wind from behind the sails of a mill.

TALTHUR The lever on the side of a post mill tailpole used to raise the ladder when turning the mill, *12.*

TENTERING GEAR General expression for the bridgetree etc. which are used for adjusting the gap between the stones.

THRIFT Handle which holds mill bills and picks, *44.*

TOLL The taking of some flour or meal in payment for grinding.

TOWER MILL A windmill with a brick or stone tower and a cap which turns to face the wind, *25.*

TRESTLE The sub-structure of a post mill.

TRIANGLES Cranks which operate the striking gear of patent sails, *32.*

TRUCK WHEEL Centres the cap of a tower or smock mill, *22.*

TRUNDLE WHEEL Similar to a lantern pinion with the top flange removed and the staves shortened, *53.*

TWIST PEG Adjusts the angle of the shoe, *42.*

UNDERDRIFT Stones driven from below, *47.*

UPRIGHT SHAFT The main shaft of a smock or tower mill which passes through several floors to drive the machinery.

VANE Alternative name for shutter or fan blade.

VAT The casing which encloses the stones (tun), *42.*

WALLOWER The bevel gear driven by the brake wheel when an upright shaft is used, *22, 53.*

WEATHER The twist of a sail, *29.*

WEATHER BEAM Alternative name for the breast beam.

WHEEL-AND-CHAIN GEAR Mechanism which turns the cap of some old smock and tower mills, *19.*

WHIP The main timber of a sail when a stock is used, *29.*

WINDING Turning the sails to face the wind.

WINDSHAFT The main axle of the sails, *12, 16.*

WIRE MACHINE Type of flour dresser which uses wire mesh to grade the meal, *49.*

Y WHEEL A wheel with Y-shaped forks around the rim which give the rope or chain increased grip, *33.*

Index of windmills mentioned in this book

Select bibliography

Batten, M. I., *English Windmills*, vol. 1.
Beatson, Robert, F.R.S.E., *Vertical and Horizontal Windmills*—An Essay.
Black and Co. Ltd., *The English Scene.*
Brangwyn, Frank, and Preston, Hayter, *Windmills.*
Clarke, Allen, *Windmill Land.*
Coles Finch, William, *Water mills and windmills.*
Darby, M. C., *The Draining of the Fens.*
Dull, William A., *The Norfolk Broads.*
Foord Hughes, A. *Windmills in Sussex.*
Freese, Stanley, *Windmills and Millwrighting.*
Farries, K. G. and Mason, M. T., *Windmills of Surrey and London.*
Harrison, H. C., *The Story of Sprowston Mill.*
Hemming, Rev. Peter, *Windmills in Sussex.*
Irving, Lawrence, *Windmills and Waterways.*
Kent County Council, *Windmills in Kent.*
Kershaw, J. C., *Wind, Tide and Stream.*
Long, George, *The Mills of Man.*
Mais, S. P. B., *England of the Windmills.*
Martin, Edward A., *Life in a Sussex Windmill.*
Paddon, J. B., *Windmills in Kent.*
Skilton, C. P., *British Windmills and Watermills.*
Smith, D., *English Windmills.* vol. 2.
Transactions of the Newcomen Society. *Various.*
Thurston Hopkins, R., *Old Watermills and Windmills.*
Thurston Hopkins, R. and Freese, Stanley, *In Search of English Windmills.*
Thurston Hopkins, R., *Old English Mills and Inns.*
Wailes, Rex, *The English Windmill.*
Wailes, Rex, *Windmills in England.*
Wolf, Alfred R., *Windmills.*
Woods, K. S., *Rural Crafts of England.*